Geology for Ground Engineering Projects

Geology for Ground Engineering Projects

CHRIS J. N. FLETCHER
Geological Consultant, United Kingdom

CRC Press
Taylor & Francis Group
Boca Raton London New York

CRC Press is an imprint of the
Taylor & Francis Group, an **informa** business

Cover photographs: The geological features presented in the collage illustrate the wide range of the geological processes that could be relevant to geological and ground models for ground engineering projects. Upper photograph - glacial deposits on the north side of Mount Qolomangma (Mount Everest), Tibet. There was approximately 3 cm lateral movement of the peak due to plate tectonic processes during the 2015 Nepal earthquake. Lower photograph - sedimentary rock sequence exposed in a cut slope at Penjom Gold Mine, Malaysia. Closely-spaced discontinuities (bedding) dominate the rock face, and tropical weathering has decomposed the upper part of the exposure. Left inset photograph - rock core sample (8 cm diameter) of granite from site investigation for tunnel, Hong Kong. Two phases of granite can be identified with a sharp, strong, fused contact. Right inset photograph - Mazier sample (8 cm diameter) of finely laminated mud and silt from site investigation for high-rise building foundations, Hong Kong. The sediment was deposited in a limestone cavity over -100 m below ground level at a time of low sea level during the last glacial period.

CRC Press
Taylor & Francis Group
6000 Broken Sound Parkway NW, Suite 300
Boca Raton, FL 33487-2742

© 2016 by Taylor & Francis Group, LLC
CRC Press is an imprint of Taylor & Francis Group, an Informa business

No claim to original U.S. Government works

Printed on acid-free paper
Version Date: 20160323

International Standard Book Number-13: 978-1-4665-8549-2 (Paperback)

Library of Congress Cataloging-in-Publication Data

Names: Fletcher, Chris J. N., 1943- author.
Title: Geology for ground engineering projects / Chris J.N. Fletcher
Description: Boca Raton : Taylor & Francis, a CRC title, part of the Taylor & Francis imprint, a
    member of the Taylor & Francis Group, the academic division of T&F Informa, plc, [2016] | Includes
    bibliographical references and index.
Identifiers: LCCN 2015036665 | ISBN 9781466585492
Subjects: LCSH: Engineering geology.
Classification: LCC TA705 .F64 2016 | DDC 624.1/51--dc23
LC record available at http://lccn.loc.gov/2015036665

Visit the Taylor & Francis Web site at
http://www.taylorandfrancis.com

and the CRC Press Web site at
http://www.crcpress.com

Contents

Preface

I came late to working on the geology of engineering projects – most of my career was involved with traditional geological mapping with occasional mineral exploration programmes thrown in. All this changed when I was appointed head of the Hong Kong Geological Survey, which is part of the Civil Engineering and Development Department of Hong Kong. The primary geological mapping had been completed and focus had been turned towards supporting and supplying data for a plethora of planned and ongoing infrastructure projects: foundations for bridges, high-rise buildings and an airport, tunnels, railway stations, quarries, cavern excavations and numerous cut slopes. Added to this, shortly after I arrived, there were two fatal landslides, which required significant geological input. Site investigation material abounds throughout Hong Kong – a region of highly varied geology over a relatively small area. Although boreholes were logged according to set guidelines, I felt there was so much more geological information to be gained from the core samples. Some rock and soil features provided additional clues for the overall geological evolution of the region, whereas on a few project sites the geology encountered was truly astounding, for example finding laminated clays within a weathered granite over 100 metres below sea level. This was not only a geological conundrum but also greatly affected the engineering works. Therefore, I have written this book not only to encourage engineers to take on a little of my enthusiasm for geology but also to help them have a constructive and informed dialog with the geologist in order to establish the best ground and design models and, thereby, reduce the risk of encountering unfavourable ground conditions during the project.

Acknowledgements

This book would not have been possible without the help and support of numerous geologists, geotechnical engineers and engineering geologists with whom I have worked over my long career in many parts of the world. In particular, the cooperation and assistance of professional staff from consulting and contracting companies, government departments and universities in Hong Kong are greatly appreciated for granting me access to site investigation material and unpublished reports over a wide spectrum of geological situations. The permission to publish their data has been of paramount importance in the compilation of this book. I am particularly grateful to the Geotechnical Engineering Office, Civil Engineering and Development Department, Hong Kong SARG, for permission to use some cited figures and photographs. The cited geological maps used in the book have been sourced from the Geological Surveys of Hong Kong, Britain and the United States. The satellite images are courtesy of Landsat, U.S. Geological Survey, together with other government sources that provided data. The provision of photographs from Bill Fitches, Andy Hanson, Keith Fletcher and Peter LeCouteur is gratefully acknowledged. I also highly appreciate the geological discussions over the years with Steve Hencher, Dick Martin, Rod Sewell, Steve Parry and Diarmad Campbell.

This book has been greatly enhanced by the willingness and interest of the authors of the case studies to take time out of their commitments to present their findings in a succinct, illustrative and informative manner. I thank Tom Berry, John Brown, Tom Casey, Paul Mellon, Seth Pollak, Dick Martin, Mike Piek, Chris Snee, Iain Betws, Dave Jameson, Nigel Wightman, Andy Mackay, Changiz Roohnavaz, Edward Russell, Howard Taylor, Katie McInnes and John Thomas for their time and enthusiasm.

Finally, I sincerely thank my wife Helen, who has been a constant companion on many geological field trips to distant and inaccessible places. Her continued encouragement and patience over the last few years were vital to the completion of this book.

Author

Chris J.N. Fletcher is a consulting geologist based in Wales. His varied geological career has taken him to many parts of the world to undertake geological mapping, mineral exploration, and ground engineering projects and research. He graduated from St Andrew University, Scotland, before going to Canada to complete an MSc at Queen's University, Kingston, Ontario, Canada, and a PhD at the University of British Columbia, Vancouver, British Columbia, Canada. His research studied phase equilibria of metamorphic minerals and the tectonic development of the Cariboo Mountains. During his time in Canada, Dr. Fletcher worked temporarily for mining companies and the Geological Survey of Canada. He joined the British Geological Survey in 1972 and worked for them for more than 25 years, initially as part of the British government's technical cooperation programme in South Korea, Bolivia and Pakistan and later in their regional office in Wales. In 1994, he went to Hong Kong as director of the Hong Kong Geological Survey, where he was also appointed as an honorary professor in the Earth Sciences Department, Hong Kong University, and the director of their Applied Geoscience Centre. Later, he managed his own geological consulting company for five years in Hong Kong, which enabled him to visit and study a wide spectrum of ground engineering projects. Whilst in Hong Kong, Dr. Fletcher also presented BSc and MSc courses on regional and applied geology at Hong Kong University and the City University of Hong Kong and also organized core logging courses on behalf of the Institution of Mining and Metallurgy (Hong Kong Branch). His interest in the research on metamorphic and tectonized rocks continued during these years, and he made several trips to China as part of The Royal Society bilateral research programme. He was given an honorary professorship at Northwest University in Xi'an, China. Many of the photographs presented in this book were taken as part of his geological work in many parts of the world, including the Canadian Rockies, the Hindu Kush of Pakistan, the Himalayas of Tibet, North and South America, China, Korea, Brazil, Malaysia and throughout Europe.

Chapter I

Introduction

The objective of this book is to provide the engineer with an effective understanding of the composition, mode of formation and architecture of rocks and soils, geological structures and alteration phenomena, as far as they might impinge on the planning, design, mitigation of risks and financial/time constraints of ground engineering projects. Project engineers need to have a sound comprehension of geological models presented to them, in particular their reliability, their limitations and, of vital importance, the assumptions made in their formulation.

In this book, rocks and soils are described in relation to the environment of their formation, highlighting the variations in composition, three-dimensional architectures and boundary relationships that can be expected within a variety of geological environments. It is not the intention of this book to set out a comprehensive account of all rocks and geological features, to act as a definitive guide to engineering geology nor to be a text on sample description in site investigations, for there are many excellent books that comprehensively cover these subjects, for example Blythe and de Freitas (1984), Waltham (2009), Norbury (2010) and Hencher (2012), and the reader will be referred in each chapter to publications that cover specific areas of geology in more detail. Rather, this book provides an insight into, and an understanding of, the most significant geological phenomena that should be considered for many engineering projects, for example rock contact relationships, weathering patterns, solution phenomena, composition of fault zones and rock discontinuities.

Geology can present a perplexing array of concepts, terminologies, processes and rock and soil architectures to many in the engineering community, and this has led at times to a lack of understanding and constructive dialog between the geologist/engineering geologist and the engineer. This is well illustrated in many geological reports, which can overemphasize geological descriptions and principles that are not directly relevant to the engineering project in hand; subdivide rocks or superficial deposits into a plethora of complex names; include too many specific ages of the rocks (e.g. Cambrian, Tertiary or isotopic ages); and amalgamate rocks and superficial deposits into named groups, formations or igneous rock associations. Unless the engineer is conversant with the details of these geological concepts and terminologies, the report can readily become unreadable, seemingly irrelevant to the needs of the project or perhaps misinterpreted. The solution to this problem is to make sure that the geologist can present the data in such a way that it is comprehensible, focused and relevant to the engineer and that the engineer becomes more familiar and confident with the most significant geological terms, phenomena and concepts. The engineer must also have a working knowledge of how the geologist has arrived at the conclusions set out in reports or presented in geological maps and cross-sections and also fully comprehend the robustness, reliability and exactitude of the original data.

The views expressed by a significant number of engineers, in particular the less experienced, on the subject of the importance of the geological component in a project (rather dismissively or flippantly suggested to the author by professionals and students over many years) may be summarized as follows:

- Geologists describe the ground in such a way that it is too complex to be effectively incorporated into geotechnical calculations.
- Most ground conditions are relatively straightforward and can be adequately modelled for geotechnical purposes.
- If the ground is not as expected in the construction phase of the project, then it can be argued that there were 'unforeseen ground conditions'.
- Much of the detail in the geological appraisal is largely irrelevant, and the collection and analysis of the geological data can be expensive and delay projects.
- Standard ground investigation procedures, sample descriptions and material testing provide all the geotechnical parameters required for engineering design.

These views are quite often founded early in the engineer's career when it was necessary to study geology as a required component of the degree curriculum. Depending on the practical experience of the lecturers in ground engineering projects and the time allocated to geology, it is not uncommon for practicing engineers to consider that the geological course was too technical, shrouded in complex hypotheses and geological terminology and not very relevant to the main body of the degree course and their future professional needs. As a consequence, the geological component of their course commonly did not prepare them adequately for the range of ground conditions that they could encounter during their working life. However, that being said, it is also true that most engineers are generally very aware that a well-founded, geological model is an essential requirement for the successful completion of many projects but still find the full comprehension of how the model was arrived at rather mystifying and conversations with geologists or engineering geologists equally testing. But the engineer must be fully aware that:

- There is a paramount need to understand and establish the geological conditions thoroughly at the start of a project.
- Although the ground conditions may be complex, it is rare that the geology encountered during a project is totally unforeseen.
- Geological investigations can largely predict ground conditions, and therefore problems can be minimized and risks reduced.

In summary, it is the objective of this book to explain the geology that is relevant to many engineering projects in such a way that it is readily comprehensible to the practicing or aspiring geotechnical engineers, thereby allowing the engineers to understand the geological model presented to them, to be conversant with the stages in the development of such a model, to realize the uncertainties in the model and to be able to converse with the geologists in an informed and responsive manner.

FRAMEWORK OF THE BOOK

Each chapter provides a brief description and classification of the rocks, superficial deposits, geological structures and weathering phenomena that are commonly encountered in engineering projects. In addition, an initial section of most chapters, titled 'Engineering considerations', that focuses on the main features that could be relevant to engineering projects, is provided. A large variety of illustrations of the different geological phenomena has been included so that the readers will be able to readily recognize and understand their significance in their own project. In particular, numerous photographs of the site investigation rock core and undisturbed soil samples are presented, as it is this type of material that the engineer will most frequently encounter, both on site or more commonly as borehole photographs in reports.

An illustration of how the different geological components of a site can impinge on the engineering design is provided by a new cut slope in Turkey (Figure 1.1). The slope consists of a wide variety of rock types, including sedimentary, volcanic and igneous intrusive rocks, discontinuities, faults and weathering features – all of which must be placed in a coherent geological framework and assessed for their geotechnical significance. Of particular concern at this site is the future stability of the slope, which could be influenced by a number of factors, including the composition and weathering characteristics of the component rock

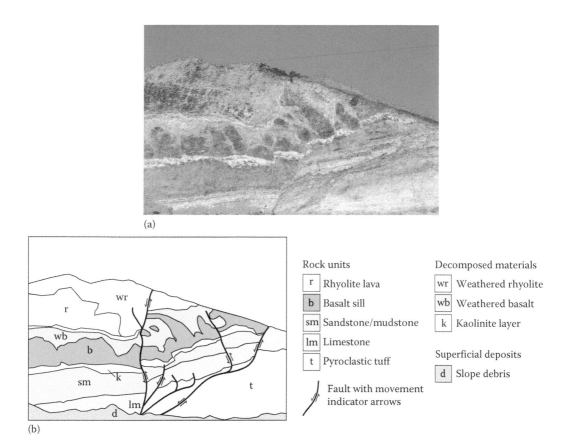

Figure 1.1 Cut slope exposed during the construction of development platforms, western Turkey. (a) Cut slope composed of a layered sequence of sedimentary (limestone, mudstone sandstone) and volcanic (tuff and basalt) rocks. Steep and low dipping faults have displaced the sequence. (b) Diagram of the geological relationships seen in the cut slope. Faults are indicated by thick continuous lines and their relative movements by arrows.

types, for example the presence of thin clay-rich layers (k) above and below a basalt sill, swelling clays in the tuffs (t), weathering profiles above the basalt (wb) and rhyolite (wr), the variability in the orientation and spacing of the rock joints through the rock sequence, the nature and orientation of the faults and the hydrogeology of the slope. This cut face provided a template for the geology and probable engineering considerations for the whole site and therefore should be taken into account when planning for additional geological surveys, site investigations, sample testing and hydrogeological studies.

An important aspect of this book is the inclusion of over 20 project case studies, which are directly relevant to the chapter in which they are placed. The key engineering significance of the geological phenomena encountered in these projects is highlighted. Case studies are presented from around the world, including Hong Kong, United States, Scotland, Wales, Malaysia, Bangladesh, Kazakhstan, Mongolia and Jordan. Thus, the case studies include a range of rock and soil associations that have been subjected to widely different weathering and erosion conditions. Many are written by the project engineers or geologists actually involved in the projects, whereas others have relied on published information.

Most chapters also include satellite images (Landsat) of specific areas as compiled by Google Earth, where particular geological features and present day environments are encountered. Location data (latitude and longitude) for these images are provided so that the reader can access and study the Google Earth data for themselves – a facility that allows both vertically exaggerated and inclined views to be generated. The knowledge of the availability, superb quality and contained information of such images will hopefully encourage the readers to use this readily accessible and important data source as a starting point for their own projects.

Rocks and soils are described with reference to the environment of their formation, thereby highlighting the variation in composition, vertical and lateral geometries and

Figure 1.2 Slope failure in weathered tuff. Po Shan Road, Hong Kong. This 1972 landslide above a housing development area resulted in 2 tower blocks being destroyed and 72 fatalities. (Published with permission of Civil Engineering and Development Department, Hong Kong Special Administrative Region.)

contact relationships that could be expected within a variety of rock associations. To understand more easily the ancient geological environments, reference is made to similar environments seen today.

Finally, the book is illustrated by a large number of photographs of materials recovered from site investigations in Hong Kong and includes several case studies from that region. This rapidly developing urban conurbation provides an ideal ground model for other parts of the world as it encompasses a comprehensive variety of geological phenomena (e.g. igneous and sedimentary rocks, tectonic structures, superficial deposits, weathering features), which were encountered during the construction of major engineering projects, including the construction of bridges, mass transit systems, drainage tunnels, high-rise buildings, dams, retaining walls and reclamations. All these projects have been overseen by the Geotechnical Engineering Office (Civil Engineering and Development Department, Hong Kong Special Administrative Region), which has provided rigorous geotechnical control of projects from the design through to construction. As a consequence of these numerous engineering projects, there are many thousands of accessible site investigation records and reports, which cover a wide range of geological situations. Hong Kong has developed a robust procedure to counter environmental hazards and has undertaken numerous forensic investigations of major landslides within the urban environment (Figure 1.2).

THE GEOLOGICAL MODEL

The geological model sometimes contained within a 'baseline survey' characterizes the geology of the project site with particular reference to the type of engineering structure under construction or ground being investigated (Fookes, 1997; Fookes et al., 2000; Bock, 2006). It forms the basis for the formulation of the 'ground model' that characterizes the bodies of rock and soil with similar engineering properties, identifies and describes the contacts between different bodies and highlights the orientation and continuity of the main discontinuities (Knill, 2002; Geotechnical Engineering Office [GEO], 2007). The subsequent 'design model' is specifically concerned with the response of the engineering works on the ground and uses the geotechnical parameters, defined in the ground model, to undertake numerical engineering analysis to determine, for example the bearing capacity for foundations or the factor of safety of cut slopes. It may be that a site could have an extremely complex geological model in terms of, for instance rocks types, fold structures or type of contact relationships, but the engineering design can be relatively simple. For example, the vertical cores presented in Figure 1.3 show firstly a contact between two granite bodies and secondly a series of sub-vertical normal faults cutting a layered limestone sequence. Both constitute important features in terms of the complex intrusive and tectonic histories of the area; however, from the design model perspective, they are of little significance; the geotechnical characteristics of the two granites are very similar and the contact between them is fused and shows no sign of alteration (Figure 1.3a), and the fault planes have been totally welded by calcite of a similar strength to the host limestone (Figure 1.3b). Although these geological features will certainly form an important component of the geological model, they will have only slight or no influence on the geotechnical behaviour of the rock mass.

Much has been written on the modelling of ground conditions related to engineering projects, but all too commonly the geology is reduced to the lowest denominator so that it can be readily incorporated into engineering calculations. Descriptions of material samples from site investigations, including rock core, soil Maziers, standard penetration test (SPT) samples, vibrocores and trial pits follow approved descriptors, but this can lead to essential

(a)   (b)

Figure 1.3 Complex geology and the design model. Examples of the core from vertical site investigation boreholes in Hong Kong where the geology is variable, but the rock masses can be considered as relatively homogeneous entities for the ground/design models. (a) Contact between two phases of granite intrusion: an early medium-grained granite and a later fine-grained granite that has intruded against and into the overlying coarser-grained granite. The contact zone is completely welded with no evidence of hydrothermal alteration nor development of structural disconti-nuities. (b) Series of vertical, normal micro-faults in thinly bedded metasiltstone and marble sequence. The fault planes have been cemented by calcite.

geological features being overlooked or in some cases ignored. Such oversights raise the risk of encountering unexpected ground conditions during the construction phase of the project, which would almost certainly increase costs and result in project overruns and ultimately could lead to expensive litigation (Fletcher, 2006).

The geological model characterizes the ground conditions of a site and focuses on the geo-logical, geomorphological and hydrogeological features that are relevant to the engineering project. It is developed using a wide variety of surveys and forms the basis of many ground applications, including foundation designs, environmental issues, and material resources (Figure 1.4). The model may be presented as geological maps, cross-sections, baseline sur-veys, written reports or even diagrams but must be as complete as possible and must be updated on a regular basis as more information is obtained through site investigation bore-holes, trial pits, geophysical surveys and site preparations. The establishment of the geologi-cal model as a requirement for any geotechnical project has been stressed as good practice by many authors and institutions (e.g. GCO, 1987; Fookes, 1997; Bock, 2006; GEO, 2006). An incomplete or poorly constrained geological model will almost certainly cause geotech-nical problems at a later stage of the project. The message is clear – a good initial geological model reduces geotechnical risk and is of paramount importance for the planning of addi-tional site investigations and the development of the ground model. In certain instances,

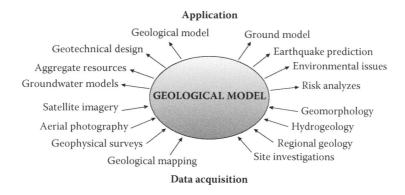

Figure 1.4 Schematic representation illustrating the central role of the geological model that is derived from a variety of sources and some of the applications that evolve from that model.

the geological model may necessitate the repositioning of the building structures in order to avoid adverse ground conditions (see Case Study 11.1), or in exceptional circumstances the abandonment of the structure completely (see Case Study 12.2). Engineering geologists, in close working relationship with other professionals, have a crucial role in the formulation of such geological models. It is one of the main objectives of the geologist to identify and anticipate problems that might arise during the duration of the project (Hencher and Daughton, 2000).

This book is aimed to assist in the development of the best geological model possible; as such it does not dwell on the minutiae of geological classification nor does it describe the complete range of rock types, tectonic structures or geological environments. More importantly, it provides details of those geological features that are directly relevant to the engineer, including fault compositions, weathering variations, contact relationships and discontinuities.

In summary, the engineer must be fully conversant with how the geological model was derived, what data was used and its reliability, and comprehend the uncertainties, and thereby the potential risks, inherent in the model. The essential elements in the development of geological and ground models should include

- Early definition of the likely geological and geotechnical complexity
- Identification of key areas of geotechnical uncertainty for further investigation
- Characterization of the ground in terms that are relevant to the engineering application
- Assessment of relevant external factors which may affect or be affected by the project
- Review of the geological and ground models during construction

Geological modelling procedures

The formulation and refinement of the geological model should essentially follow an ordered series of tasks and be an integral part of the project, not an adjunct. The following list of tasks is not comprehensive as additional surveying techniques (e.g. geophysical surveys) can be incorporated should the situation require the following:

1. Check the site location with respect to the regional geological setting, plate tectonic boundaries, hazards (seismic, volcanic, etc.), major topographic features and climate characteristics. It is useful to view satellite images of the site.

2. View all published geological maps at all available scales, particularly those at large scale; check with the organization that undertook the mapping to ascertain whether there is any additional information available since the last survey.
3. Collate all available data from site investigations in the vicinity of the site, assessable databases and other sources – the desk study.
4. Undertake an aerial photograph investigation of the site using historical photographs where available in order to establish main geomorphological and hydrogeological features and developmental history of the site.
5. Carry out a walk-over survey of the site to ground check previous observations and note any unrecorded relevant data.
6. Complete a preliminary geological map of the site and report on the main rock types, superficial deposits, faults, discontinuities and weathering characteristics (among others). Draw cross-sections, block diagrams and sketches to augment the map and report.
7. Identify the geological and geotechnical problems, uncertainties and environmental risks.
8. Make recommendations for a ground investigation that focuses on the geological uncertainties, thereby assuring that the ground model is as robust as possible.

The development of the geological model is an ongoing and iterative process. As more information is acquired, the model needs to be reassessed and refined. This is partially important once the site investigation data becomes available. The observation of Fookes (1997) makes clear the importance of a systematic approach to the development of the geological model: *"...an experienced, competent engineering geologist should, from a desk study, be able to anticipate 50% of the potential geological problems at a site. With a walk over survey this should increase to 65%. With a site investigation 95% of potential problems should be capable of being identified."*

SITE INVESTIGATION SAMPLES

The photographs of the site investigation samples that are presented in this book are mainly of rock core recovered during rotary drilling and weak soil collected in plastic Mazier tubes during the drilling process or in small diameter metal liners used during standard penetration tests (SPT). Details of the equipment used in the recovery of these samples are described by Hencher (2012, pp. 158–165).

For the majority of engineers, it is the borehole logs and photographs of the core that will form their main knowledge base for the ground conditions of the project; in many instances, the engineer does not actually inspect the rock and soil samples or question the validity or completeness of the borehole logs. It should be stressed, however, that borehole logs often form the basis for contractual disputes, and therefore they must be as accurate and comprehensive as possible with respect to their use in geotechnical design. The engineer is totally reliant on the rigor, skill and experience of the logging geologist but needs to be aware that logging is undertaken using a prescriptive set of parameters and formats that are outlined in publications such as Geotechnical Control Office (GCO) (1988), British Standards Institution (1999, 2002) and Norbury (2010). However, such descriptive procedures can lead to vitally important geological features being overlooked, for example the soil and rock is described in simplest terms, the use of general descriptors for variable material, the characteristics, spacings and orientations of joints may be listed without reference to joint sets (Hencher, 2012) and the lack of detailed description of important geological features, particularly those of critical importance for the geological model. It is worth remembering that most borehole logs are purely factual without any reference to geological process or type of engineering project. Accurate, comprehensive and understandable descriptions of borehole material are often more helpful and informative than a plethora of unconstrained index test results.

The function of a borehole log is to assist the geologist and engineer in the creation of the geological and ground models through the description of the rock and soil types, identification of discontinuities, classification of the weathering (decomposition) grades, as well as providing initial estimates of the material strengths, details of the total core recovery (TCR), solid core recovery (SCR), rock quality designation (RQD) and fracture index (FI) (Geotechnical Control Office, 1988; British Standards Institution, 1999; Norbury, 2010). The description of the soil also follows prescribed formats, such as the weathering classifications of granite (Geotechnical Control Office, 1988) and mudstone (British Standards Institution, 1999); however, local classifications of the weathering characteristics of other rock types are valid as long as the system is fully documented. Logs of soil are commonly extrapolated from the inspection of the ends of the Mazier tubes; it is the exception that the Mazier tubes are cut open and logged in their entirety. Opened Mazier tubes may provide important additional information (see Case Study 12.2).

As an illustration where core description, following recognized procedures, can omit mentioning features that are important components of the geological model is presented in Figure 7.18. The core is from a site investigation in Hong Kong and was simply recorded as a narrow quartz vein in siltstone. However, this section of core provides essential information as to the complex, tectonic history of the site indicated by three generations of quartz veining that developed under varying stress regimes within a shear zone. It also strongly suggests that other shear zones, possibly with the same orientation, will be present close by and these could contain weak fault material in addition to hard quartz lenses and veins. Such shear zones could also act as aquicludes and may, in certain topographic situations, act as release surfaces for landslides. With this additional knowledge not shown in the core logs, subsequent ground investigations could be better targeted to locate and characterize such shear zones.

In summary, a borehole log by a competent geological practitioner should

- Factually record the material in a systematic and accepted manner such that there is constancy between logs, no matter who has logged the samples
- Be sufficiently detailed as to enable the engineer to visualize the ground conditions but not over complicated so as to obscure the geotechnical considerations
- Provide limited interpretation of the material and structures if relevant to the engineering properties of the rock mass
- Identify geological phenomena that could be important in the development of both the geological and ground models

Chapter 2

Geological fundamentals

In this chapter, the reader is introduced to the fundamental concepts behind geological modelling with special reference on how they can impinge on ground engineering projects. The engineer should be acquainted with, and have a basic understanding of, geological time, unconformities, global geological evolution, the formulation of geological histories and the definition of terms commonly used to describe rock and superficial sequences as shown on geological maps and in reports. References to important scientific papers and textbooks that provide more information on these subjects are provided throughout this book.

The foundations of modern geology were established in the late eighteenth century when it was recognized that the time period required to form the Earth must have been considerably longer than the religion-based estimates of a few thousand years. James Hutton in 1787 surmized that the geological structure he saw near Lochranza, western Scotland, must have evolved over many millions of years. There, he observed that the steeply dipping rock strata exposed along the foreshore for over a kilometre were sharply overlain by a completely different sequence of near-horizontal rock strata (Figure 2.1). He proposed that the formation of these rocks comprised distinct evolutionary stages: sedimentation, tilting, uplift, erosion and finally, a further period of sedimentation at a much later time. In effect, he had established a robust geological history that was consistent with his observations and comprised a rational sequence of geological events. From other observations throughout Scotland and elsewhere, he concluded that the rates of sedimentation and erosion at the Earth's surface have been constant over geological time, and igneous rocks were produced by the cooling of molten rock. Hutton's pioneering work laid the foundation for modern geological thought that was later developed by Charles Lyell. Their publications were highly influential and became a key component of Charles Darwin's 'Origin of Species'.

Another major step in the development of geological thought was the realization that continents had moved across the Earth's surface over vast periods of time – a proposal that has been put forward but rejected by many since the mid-nineteenth century. It was not until the 1960s that evidence for sea-floor spreading provided the evidence that continental drift was a scientific reality (Vine and Matthews, 1963), thereby heralding the present-day perspective on the global evolution of the Earth's crust and insights into, and explanation for, past geological processes.

ENGINEERING CONSIDERATIONS

For the geologist to provide the most robust, comprehensive and sound geological model, it is essential that all the processes and the relative timing of those processes are considered and thoroughly understood. It may be that the thought behind the geological model is not obvious in the final presentation, for example in the form of maps, reports, baseline surveys and

Figure 2.1 'Hutton's Unconformity' at Lochranza, Isle of Aran, Western Scotland. Steeply dipping Silurian schists are overlain with angular unconformity by sub-horizontal sandstones of Carboniferous age. The contact between the two sedimentary sequences therefore represents a time interval of at least 50 million years – a period of uplift (mountain building) and erosion.

cross-sections. However, without the scaffolding that backs up these presentations, the reliability of the overall geological model could be misplaced and certainly will be questioned should the model be challenged in arbitration or court proceedings related to claims. Obviously, it is the responsibility of the geologist or engineering geologist to provide the best geological model possible, but the engineer should be fully aware of any assumptions that are inherent in the model, the possibility of alternative models, the exactness of the geological linework, the important components of the timeline of the geological processes that are reflected in the distribution of the rock and soil masses and the relative importance and continuity of discontinuities. It is essential that the engineer has a basic understanding of the consequences of the geological model even though the details and background are not fully appreciated. The engineer must also be confident enough to comprehend the components of the geological model and be able to question and discuss its reliability, consistency and rationality.

The fundamental concepts of geology with which the engineer should be familiar are geological time, global plate tectonic evolution, sea level rise and the processes of formation of main rock types, superficial deposits and weathered materials, and the generation of discontinuities in rock and soil. The first three of these will be explored briefly in this chapter, whereas the remainder are included in the later chapters of this book. In summary, the engineer should be able to

- Converse confidently with a geologist by having a working knowledge of the fundamental geological principles.
- Comprehend the geological model in its entirety, although the scientific details behind the model may be less thoroughly understood.
- Appreciate the limitations, reliability and inherent assumptions of the geological model.

Four case studies that illustrate the importance of understanding basic geological principles have been included in this chapter. The first describes the offshore foundation conditions that were encountered during the construction of the reclamation for the Hong Kong Airport (Case Study 2.1). The sequence of Pleistocene and recent sediments reflect on the changes in the worldwide sea level that has been determined by the climate change and the resultant growth and melting of the polar ice caps. The geological model was essential for the foundation designs of runways and infrastructure developments, locating subsurface sand resources for reclamation works and establishing the geometry of the marine mud that needed to be dredged. The three other case studies do not relate directly to engineering projects but describe road cuts in sequences of sedimentary rocks separated by unconformities (Case Study 2.2) and metamorphic and intrusive igneous rocks (Case Study 2.3), and a natural exposure in weathered and faulted volcanic rocks (Case Study 2.4). In each of these cases, the possible engineering aspects of constituent geological features seen in these exposures and their geological histories are considered.

GEOLOGICAL TIME

The geological timescale has developed over the last 200 years from the initial recognition of a regular distribution of distinct sequences of sedimentary rocks with different fossil assemblages to the erection of named geological eras and periods. The divisions of the stratigraphic column (Figure 2.2) are recognized throughout the world and were named after

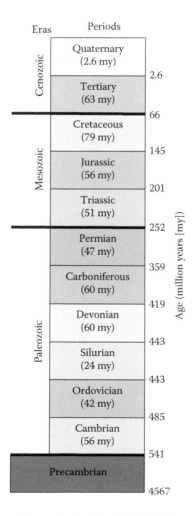

Figure 2.2 Geological time scale according to Harland et al. (1982). The exact ages of some of the boundaries are slightly modified as more data have become available.

the regions where they were first described, for example the Cambrian Period was named after the Cambrian Mountain of Wales and the Jurassic Period after the Jura Mountains of the Western Alps. Absolute dates of the boundaries between the divisions are defined by the use of radiometric techniques, for example the K/Ar, Sr/Rb, U/Pb and ^{14}C methods that are dependent on the relative concentrations of the naturally occurring radioactive isotopes compared with their decay products (Faure, 1986). The first three methods are used mainly on igneous rocks, which provide an age of the crystallization of the constituent minerals from the molten rock. Where the igneous rock, for example a lava flow or tuff layer, is interbedded with sedimentary rocks, the age of the sedimentary sequence may be inferred. The ^{14}C isotopic method is used on carbon-bearing material such as wood, bone or shell; however, the maximum age that can be measured using this method is about 40,000 years and therefore only applicable to the most recent deposits.

Isotopic dating provides a valuable tool in the dating of intrusive igneous rock bodies, which previously could only be given a maximum date based on the ages of the country rocks they intrude. Using these methods, it has been possible to group igneous bodies into different intrusive suites that have similar ages and also provide evidence for the geochemical evolution of the magma.

For the geotechnical engineer, the geological age of a rock as indicated by named eras and periods or absolute isotopic ages may seem to be of little importance to their investigations, even though frequently included in geological reports. To a large extent, this is true, for example whether a rock is of Devonian or Jurassic age is generally irrelevant for engineering purposes, for there can be just as much variation in the geotechnical properties of rocks belonging to a particular period as between periods. Nevertheless, although the age of a rock may be of little significance to the engineer, it will be an important factor in the establishment of the geological history of the project area, provide additional information on the rocks and soils through regional correlations, form an essential component of geological maps and cross-sections and can be an important basis of the geological model.

Unconformities

An unconformity is one of the most important features in the geological history of an area for it represents a break in the continuity of deposition within a sequence of sedimentary rocks (Figure 2.1) or a break between the uppermost part of an eroded igneous body and the overlying sedimentary strata. The unconformity implies a considerable time gap in the geological evolution of the rocks when uplift, folding and erosion of the older rocks took place before the deposition of the younger rocks. This time gap can be many tens of millions of years long, and therefore evidence for what happened during that time has to be inferred from the structure of the underlying rocks or by reference to other areas. Most unconformities, therefore, represent a period of time when the area was land and subjected to weathering and erosion, prior to regional subsidence or sea level rise that renewed sediment deposition. Alternatively, in more restricted areas, the accumulation of younger sediments may take place in topographically low areas, for example in lake beds and river valleys, or adjacent to steep hillsides or cliffs. In many cases, the surface of unconformity is overlain by coarse-grained sediments, for example conglomerates, that were deposited in high-energy environments as water level rose or the ancient land mass subsided.

The classical angular unconformity displays folded strata that are abruptly truncated by commonly flat lying younger sedimentary rocks. In South Wales, folded Carboniferous limestones are overlain by a thick sequence of Jurassic calcareous sandstones that were deposited close to an ancient shoreline (Figure 2.3a). This surface of the unconformity, which is planar

(a)

(b)

(c)

Figure 2.3 Unconformities. (a) Angular unconformity between steeply dipping Carboniferous limestone strata and sub-horizontal Jurassic calcareous sandstones. The surface of unconformity at this locality is approximately planar, Dunraven Bay, South Wales. (b) Unconformity between Palaeozoic folded strata (dark grey) and horizontal Quaternary lake sedimentary rocks (light brown), southern Tibet. The surface of unconformity is defined by a rugged, lake floor topography. (c) Triassic breccias cascading over horizontal Carboniferous limestone beds, Ogmore, South Wales.

at this locality, is therefore a gap in the geological record of approximately 100 million years when the original fossilerous calcareous sediments were buried, lithified, folded, uplifted and eroded prior to the deposition of the overlying sediments (Wilson et al., 1990). A more irregular surface of unconformity can be traced across the landscape in southern Tibet where folded Palaeozoic strata are overlain by a thick sequence of mudstones, sandstones and conglomerates that accumulated in a Quaternary lake behind the uplifting Himalaya mountain chain (Figure 2.3b). Rarely, steep eroded cliffs of a former shoreline are blanketed by more recent sediments. An example of this is seen in Wales (Figure 2.3c) where Triassic sedimentary breccias containing an assortment of angular blocks accumulated against an ancient stepped cliff of horizontal Carboniferous limestone. The same unconformity is shown on the geological map of the Mendips, South West England (Figure 3.3). In some areas, multiple unconformities can be recognized, even within a restricted area. For example, in southern Portugal, folded Carboniferous sandstones are abutted by sub-horizontal Jurassic sedimentary breccias and conglomerates. Both are overlain by cemented, Quaternary alluvial gravels (see Case Study 2.2).

Some unconformities can be identified on geological maps, for example where the surface of unconformity has an irregular expression on the topographic surface and the bedding traces above and below the unconformity are significantly different (see Figure 3.3). However, in many places, the unconformity is not evident on the geological map and is only identified by reference to the legend or stratigraphic column that accompanies the map.

From a geotechnical engineering perspective, unconformities can form a major discontinuity across which the rock types and their structure change radically. They may also influence the hydrogeology of an area or be associated with a weathered soil horizon. However, their significance is paramount for the formulation of the geological model as they establish the geological history of an area, thereby allowing the distribution of the rocks to be more fully understood and predicted (see Case Study 2.2).

PLATE TECTONICS

In the 1960s, the theories of sea-floor spreading and continental drift under the umbrella term plate tectonics were developed and tested. It was a major leap forward in the understanding of the evolution of the Earth's crust and the underlying mantle and provided explanations for the formation and distribution of the oceans, continents, mountain chains, island arcs, volcanoes and earthquakes, as well as the correlation of regional geological features seen across the world. It now forms the basis for the understanding of past geological regimes and allows the global relationships between the rock types and tectonic structures to be comprehended and logically explained.

This section provides a brief explanation of the major plate tectonic settings to be found around the globe, most of which can be readily seen and studied by reference to satellite images. The reader is referred to the more comprehensive accounts of plate tectonic theory as presented in other publications (e.g. Anderson, 1989).

The Earth's thin, outermost layer, commonly referred to as the *crust*, is composed of either silica-rich, sedimentary rocks, granites and metamorphic rocks (*continental crust*) or iron/magnesium-rich volcanic rocks (*oceanic crust*). The thickness of the continental crust varies from over 36 km beneath some mountain chains such as the Himalaya to less than 12 km in the central parts of some continents. The oceanic crust is much thinner and on average only 6 km thick. The crust 'floats' on the much denser mantle material.

The plate tectonic theory proposes that the Earth's crust is divided into 7 major plates and some 12 minor plates (Figure 2.4), and these move relative to one another over the surface of the Earth, a process known as *continental drift*. The plates move at an average rate of 70 mm/year but can attain rates of up to 100 mm/year. Three types of active plate boundary are recognized on the Earth's surface:

- *Divergent boundary*: Where new volcanic crust is formed along oceanic ridges, such as the Mid-Atlantic Ridge (Figure 2.5a and b), the ocean floor spreads away from the boundary. Here, the plates increase in size; therefore, the boundary is constructive.
- *Convergent boundary*: Where two plates move towards each other, the oceanic plate dips below a continental or oceanic plate along a *subduction zone* (e.g. western coast of South America). As the plate descends, it is partially melted, and magma so formed rises through the crust and erupts along a volcanic island arc within the uppermost plate, for example, the Aleutian Islands of Alaska (Figure 2.5c and d). Here, the Pacific oceanic plate decreases in size; therefore, the boundary is destructive.

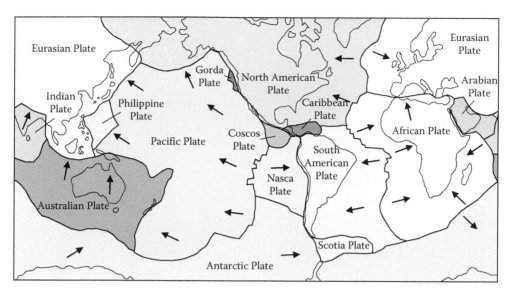

Figure 2.4 Map of the main tectonic plates and their movement vectors.

- *Transform boundary*: Where two plates slide past each other along a strike-slip fault, for example, the Chaman fault between the Afghanistan and Indian Plates (Figure 8.6a and b), where there is no consumption or creation of plate material.

The relative movement of plates also creates other structures:

- *Collision zone*: At a convergent boundary, there may be a collision between two continental plates after the ocean floor that originally separated them has been almost totally consumed along a subduction zone. Shortening of the crust during continental collision is associated with faulting, folding, igneous intrusion and metamorphism. Within the collision zone, remnants of the ancient ocean floor, island arc volcanic rocks and the marine sedimentary rocks are preserved along sutures. The best example of a collision zone is the Himalaya that were formed by the progressive northward movement of the Indian Plate and its collision with the Eurasian Plate (Figure 2.5e and f) – a movement that continues today.
- *Rift valleys*: Within continental plates, the crust can start to split apart along fault lines that represent the initial stages of the formation of a spreading oceanic ridge. A fault-bounded rift valley is formed that is progressively filled with sediments, and where the extension continues today, active volcanoes are formed as a result of upwelling magma along the weakness in the crust (Figure 8.4a and b).

Plate tectonic theories may seem rather academic to the engineer working on a small project site; however, the mechanisms of plate tectonics are the engine behind geological evolution. Thus, the plate tectonic setting of a site should be known at the start of any project – a cursory acknowledgement for some geological environments and types of engineering project, for example building foundations in a weathered granite located in a stable continental area, but a major consideration in others, for example the alignment of a road or tunnel in

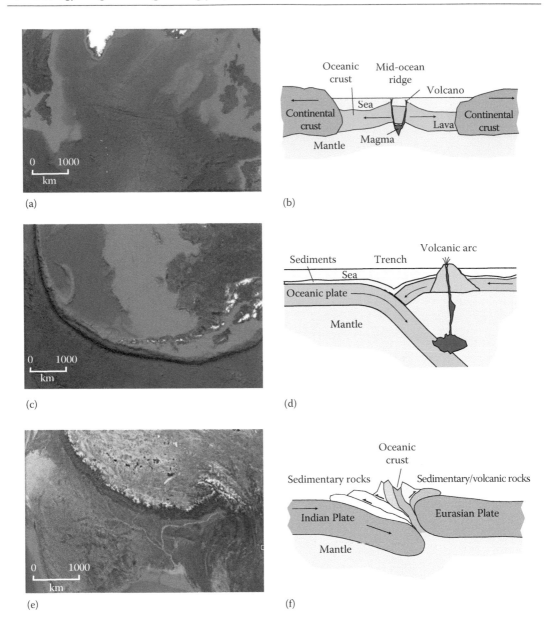

Figure 2.5 Examples of tectonic plate boundaries: (a) Google Earth, Landsat image of the Mid-Atlantic Ridge that passes through Iceland – a divergent plate boundary separating the North American and Eurasian continental plates. (b) Schematic diagram of the North Atlantic showing a spreading ridge, generation of oceanic crust (mainly basalt lavas) from active volcanoes and the divergent movement of the two continents. (c) Google Earth, Landsat image of the Aleutian archipelago where the Pacific Plate is subducted northwards beneath the North American Plate – a convergent plate boundary. (d) Schematic diagram of the subduction zone and the formation of the volcanic island arc. (e) Google Earth, Landsat image of the Himalaya Mountain chain – a collision zone between the northward moving Indian Plate against the Eurasian Plate. (f) Schematic diagram of the collision zone showing remnants of the ancient ocean floor that once separated the two continents being exposed along the suture zone. The Indian Plate continues to be pushed beneath the Eurasian Plate. Sediments deposited within the ancient ocean and close to the two shore lines are uplifted and deformed by the collision.

a tectonically active area close to a plate boundary. However, in all cases, the earthquake, volcanic and geomorphic risks as defined by the plate tectonic environment must be established and considered, even if those risks are minimal. Most importantly, if the engineer is unfamiliar with the geological environment of a new project area, then the starting point for the investigation should be to establish its plate tectonic setting as it relates to potential geotechnical risks.

SEA LEVEL CHANGES

The sea level change over the last 11,700 years BP (Holocene Period), from approximately 120 m below present sea level at the end of the last Ice Age, continues today at about a global average of 3 mm/year (Martinson et al., 1987). Over the previous 350,000 years, the sea level has fluctuated in response to the retreat and advance of the polar ice caps. Three major glacial periods are separated by interglacial warm periods, when sea levels exceeded present-day levels (Figure 2.6). The melting of the ice sheets was in response to cyclical global warming episodes, the most recent of which follows the pattern of the previous ones. The intense debate that is being conducted by the scientific world and governments today concerns the extent to which humans have exacerbated the situation and whether sea levels will continue to rise at the same rates (for a summary, see Henson, 2006). Although it is commonly stated that there is a consensus that global warming is in large part the result of human activity, there are those that consider the phenomenon is largely natural. It is not the intention of this book to summarize all the arguments for and against these theses but to emphasize that sea

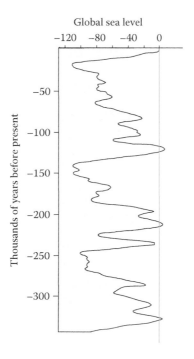

Figure 2.6 Sea level changes over the last 350,000 years, based on oxygen isotope measurements. Note that in previous three interglacial periods, the sea levels surpassed present-day levels. (After Shackelton, N.J., *Quater. Sci. Rev.*, 6, 183, 1987; Martinson, D.G. et al., *Quater. Res.*, 27, 1, 1987.)

level will inevitably rise over the next few decades and engineers must be fully aware of this trend and design projects appropriately.

Relative sea level has been lowered in some places. In Scotland, there are a series of wave-cut platforms that reach about 30 m above present mean sea level (see Case Study 6.2). This is due to isostatic rebound, caused by the unloading and uplift of the land mass following the melting of the ice cap that once covered upland regions. This uplift is greater than the global rise in sea level after the last glacial period. The lowest of the sea-cut platforms around the coast of Scotland is approximately 8 m above present mean sea level and is well preserved on the Isle of Arran (Figure 2.7c) where it is backed by an old cliff line. In Thailand, old wave-cut nips that formed between high and low tide levels are indicative of the lowering of the relative sea level (Figure 2.7d). There, the old nip is some 2.5 m above the one that is forming today. The apparent lowering of sea level in this region is due to tectonic uplift of fault-bounded blocks in the region, rather than the global lowering of sea level.

(a)

(b)

(c)

(d)

Figure 2.7 Evidence for recent sea level change. (a) Detail of a submerged forest with stumps and roots of trees set in a peat layer that is underlain by a light grey, clay layer, Borth, West Wales. (b) Remnants of a submerged forest exposed on the Borth beach at low tide. Sea defences currently being constructed include rock barriers against storm surges and offshore reefs to dissipate the force of the waves. (c) Raised beach approximately 8 m above present-day mean sea level and ancient cliff line, Lochranza, Scotland. (d) Old concave wave-cut nip in a limestone cliff, some 2.5 m above the presently forming nip, Phuket, Thailand.

An example of sea level change and the engineering solution to this problem is seen along the foreshore at Borth, West Wales. There, the remnants of an ancient submerged forest with tree stumps and roots are exposed between high and low tide levels (Figure 2.7a). The trees have been dated at about 3500 years BP and can be found at the other localities around Wales. The rise in sea level since that time swamped and killed the forest. The nearby coastal village of Borth has been flooded by the sea on a nearly annual basis for many years; as a result extensive engineered defences have been built, including rock storm barriers and multipurpose offshore reefs to combat storm surges and break up the force of the waves (Figure 2.7b). It is anticipated that the scheme will combat the immediate flooding problems of the village, but it has been designated as an 'at risk' community and future planning decisions may include the abandonment of the village.

GEOLOGICAL HISTORIES

One of the key elements in any geological investigation is the formulation of a comprehensive geological history in which the stages of formation of the various rock types, superficial deposits, deformation structures, alteration and metamorphic features and weathering phenomena are described and set out in chronological order. Without such an analysis of the geological history, it is virtually impossible to make geological maps, provide robust geological models or be able to extend one's knowledge base into areas with little or no information. It is imperative that the geological history that is erected is consistent, logical and takes into account all known information. It is also necessary to realize that geological histories may have to be amended over time as more information, such as borehole data, geophysical survey results and age dates, becomes available. It is possible to establish geological histories not only on a regional scale but also on an outcrop, core sample and even thin section scale. However, within a particular area, the geological histories, defined at whatever scale, are interdependent and must be compatible with one another. If there are inconsistencies between geological histories across an area, the overall history needs reappraisal, or possibly one has moved into a different geological regime across a major fault, for example. In the field or on site, the geological features found in rock outcrops, in borehole materials and in superficial sections should always be referenced to the geological history which has been established previously. They should either conform to that geological history or be at odds with it. In the latter case, the geological history needs to be reassessed and adjusted taking into account the new information. The formulation of geological histories may seem rather academic to the engineer, but their precision, reliability and robustness are an essential component of the preparation of the geological model. Geologists will always automatically consider the geological history of any outcrop, borehole or area during any geological survey. Any report, cross-section, map or model that does not have a consistent geological history, or the contact relationships do not conform to that geological history, are by necessity suspect and need to be reassessed and modified.

Case studies of three outcrops, which describe their component geological features and itemises their histories of formation, are provided. Although these case studies are not part of any engineering project, they introduce the key observations that may be of geotechnical significance in nearby areas. They illustrate the importance of assessing all available geological information in an area as an integral part of any engineering project. The first case study, from Portugal, describes an outcrop of sedimentary rocks that belong to different ages, depositional environments and compositions (Case Study 2.2). Two unconformities divide the outcrop into its component parts. The second case study, from Malaysia, presents the contact relationships between a highly tectonized schist and a younger granite intrusion (Case Study 2.3). The third case study is from Hong Kong and describes the phenomena related to the tropical weathering of a tuff and the deposition of an overlying layer of colluvium (Case Study 2.4).

STRATIGRAPHIC NOMENCLATURE

The use of named divisions of the stratigraphic column is of paramount importance in geological maps and reports, for it provides a readily understandable and universally accepted framework for the description of the geology of an area. The use of these divisions also allows correlation with other known rock sequences, thereby providing information as to the possible variations in rock types and their 3D form that have been encountered in the same division elsewhere. The comprehensive rules by which sedimentary sequences are divided and named are presented in international treatises, such as the North American Stratigraphic Code (2005). However, for the engineer, only the essential concepts need to be understood.

The basic unit of classification for a layered sequence of sedimentary rocks is the *formation*, which is characterized by regular distribution of rock types that can be mapped at surface or traced in the subsurface. Thus, they are defined by their lithological composition, without reference to their age, although many formations are confined to particular time intervals. Other characteristics of a formation may include its fossil content, sedimentary textures, chemistry or seismic signatures. A formation can vary from a few metres in thickness to several thousand of metres. It is commonly identified on a geological map and in the accompanying reports by the type locality named after a geographic feature such as a village, river or mountain and the dominant rock type – for example the Nantmel Mudstone

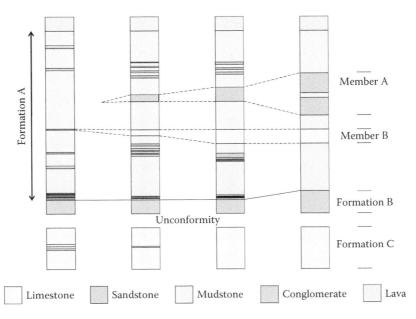

Figure 2.8 Schematic representation of a hypothetical series of four rock sequences and their suggested stratigraphic subdivision into formations and members across the area.

Formation. However, this does not imply that the formation is totally composed of mudstone, as it may contain subordinate thicknesses of other rock types such as sandstone and limestone. The description of the formation should include detailed measurements of the rock layers at the type locality where it is best exposed and an indication of the range of lithological variation within the formation.

In geological literature and on geological maps, formations may be combined into named *groups* that have a common heritage, for example a group may contain formations that were deposited in a similar depositional environment or may be bounded by unconformities or the extrusion of lava. Parts of a formation can be distinguished as non-formal *members* that may or may not be able to be mapped over an area. Figure 2.8 provides a schematic representation of a hypothetical division of a sequence of rock strata across an area.

CASE STUDY 2.1   OFFSHORE SUPERFICIAL DEPOSITS: RECLAMATION FOR HONG KONG AIRPORT

The Hong Kong Airport is located on Lantau Island to the southeast of Hong Kong (Figure CS2.1.1) and was opened in 1997. It was built on marine reclamation and two granite islands and covers an area of approximately 12 km². A decade of planning and site investigation, including closely spaced, seismic reflection surveys, preceded the construction phase of the project. A key element in the site investigation was the correlation of marine seismic surveys and the detailed analysis of some 200 boreholes of superficial deposits (Plant et al., 1998; Pinches et al., 2000). These studies enabled the 3D form of the compressible deposits to be mapped and the drainage pathways to be located. Nearly 70 million m³ of marine mud was removed from the seabed, and about 100 million m³ of material, including locally sourced, dredged sand and rock derived from the levelling of the islands, was used to form the reclamation platform.

GEOLOGY

The Quaternary sediments of the offshore areas around Hong Kong accumulated, during the last Ice Age, some 20,000 years ago, when sea level was about 120 m lower than today. A wide

Figure CS2.1.1 Hong Kong Airport during construction viewed from the southeast.

alluvial plain covered the lower reaches of the precursor to the Pearl River, and the site of Hong Kong Airport is situated over one of its tributary rivers (Figure CS2.1.2).

The reclamation covers a wide variety of superficial deposits comprising a lower alluvium unit that covers weathered bedrock and an upper marine unit (Figure CS2.1.3). The alluvium is 10–30 m thick and consists of a discontinuous basal sand and gravel layer (Figure CS2.1.4a) that is overlain by a sequence of heterogeneous silty clays (Figure CS2.1.4b). These are cut by several steep palaeo-channels, which are filled with mainly sand and silt. Within and at the top of alluvium are several distinct, intensely oxidized and desiccated layers (Figure CS2.1.4c) that were formed by subaerial exposure of sediments during times of low sea level. The clays in these desiccated layers are considerably firmer than the bulk of alluvial deposits. The alluvium is overlain by a 10–15 m thick blanket of marine mud, which consists of very soft,

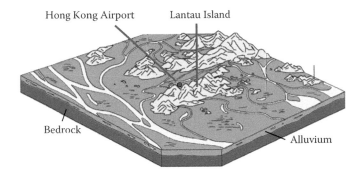

Figure CS2.1.2 Schematic representation of the Pearl River delta area 20,000 years ago. (After Fyfe et al., 2000, Published with the permission of Civil Engineering and Development Department, Hong Kong Special Administrative Region.)

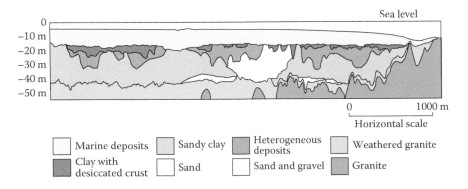

Figure CS2.1.3 WSW–ENE geological cross-section along the northern runway of Hong Kong Airport prior to the dredging of the marine deposits and formation of the reclamation. This section was based on the correlation of the marine seismic surveys with the borehole information. Note vertical exaggeration of the section. (Published with permission of Civil Engineering and Development Department, Hong Kong Special Administrative Region.)

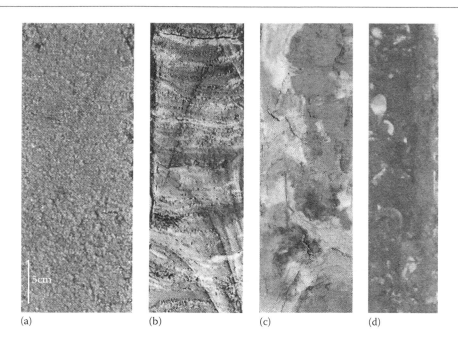

Figure CS2.1.4 Vibrocore samples of the superficial deposits beneath the reclamation at Hong Kong Airport. (a) Alluvium – loose medium sand. (b) Alluvium – thinly laminated grey silty clay and light brown fine to medium sand. (c) Desiccated crust in alluvium – mottled reddish brown and light grey silty clay. (d) Marine deposit – grey silty sand with abundant shell fragments.

greenish-grey silty clays containing shell fragments (Figure CS2.1.4d) and undisturbed grey clay with complete shells (Figure 13.7a). This unit was deposited during the flooding of the ancient alluvial plain associated with a rise in sea level, which was caused by the melting of the polar ice caps.

A major aspect of the engineering design for the reclamation was the determination of the settlement and dewatering characteristics of different superficial deposits. The final design involved the removal of the marine mud, and over most of the site, the firm desiccated crust at the top of the underlying alluvium was taken as the founding level.

ENGINEERING CONSIDERATIONS

- Composition and distribution of the superficial deposits are highly variable.
- Channel-fill deposits that cut into the alluvium are highly compressible.
- Drainage channel pathways are mainly defined by upper desiccated layers and basal sand and gravel deposits.
- Marine mud was very soft and compressible and had to be removed.

CASE STUDY 2.2   SEDIMENTARY ROCKS AND UNCONFORMITIES, TAVIRA, PORTUGAL

[37°9.43′N 7°40.56′W]

This road side exposure is composed of three sequences of sedimentary rocks separated by unconformities, each with distinctive environments of deposition, composition and geotechnical properties (Figure CS2.2.1a). The oldest rocks are planar-bedded sandstones interspersed with a few thin mudstone layers (Carboniferous). At this locality, the strata have moderate dips, although only a short distance away they have been folded to the vertical. The sandstones are medium grained for the most part but become slightly finer grained towards the top of individual beds. The surface colour of the sandstone is reddish-brown, whereas on the broken surfaces, the colour is light brownish-grey. A wedge-shaped, sedimentary breccia (Triassic) abuts the underlying sandstones. It is composed of a chaotic arrangement of angular blocks of sandstone set in a coarse-grained sandstone matrix. There is an ill-defined, sub-horizontal stratification

(a)

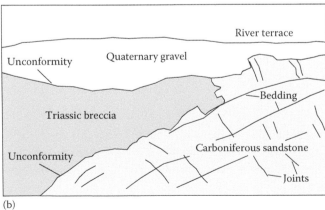

(b)

Figure CS2.2.1 Sedimentary rocks, Tavira, Portugal. (a) Road cut displaying multiple unconformities. (b) Geological sketch of the outcrop.

to the breccia with the larger blocks being concentrated in an upper layer. A scalloped surface defines the base of a thin sequence of weakly cemented gravels (Quaternary) that form a flattish terrace some 20 m above the nearby river valley. The pebbles in the gravels are subrounded and smooth indicative of abrasion in a high-energy moving water environment. The key features of this outcrop are presented in Figure CS2.2.1b.

GEOLOGICAL HISTORY

The sequential events (from 1, oldest, to 10, youngest) that make up the geological history of this outcrop are as follows:

1. Deposition of turbidite sandstones (see Chapter 4 for sedimentary environment) on the floor of a deep ocean during the Carboniferous
2. Burial and lithification of the sediment
3. Major tectonic deformation (mountain building associated with plate collision) resulting in weak metamorphism, folding, faulting and jointing
4. Uplift and erosion with the formation of the oldest surface of unconformity, which represents a 50-million-year time gap
5. Deposition of sedimentary breccias by flash floods in a dominantly arid climate on top of an underlying rock topography during the Triassic, for example red iron staining of the underlying rocks close to the ground surface
6. Burial and lithification of the breccias
7. Uplift and erosion with the formation of the youngest surface of unconformity, which represents a 200-million-year time gap
8. Deposition of river gravel within a recent drainage system during the Quaternary
9. Weak cementation of the gravels
10. Downcutting of the old gravels by the river to its present level, leaving an elevated river terrace

ENGINEERING CONSIDERATIONS

- Highly variable strengths of the different sedimentary rocks.
- Sedimentary breccia composed of hard rock fragments set in a weaker matrix.
- Tabular, lenticular and scalloped shapes to the sedimentary rock bodies.
- Major discontinuities (bedding and joints), which may affect slope stability in other areas, predominant in the sandstones.
- The oldest unconformity surface is very strong and in places inclined; as a result, it could cause deformation of driven piles.
- The terrace gravels although weakly cemented may act as aquifers at other localities.

CASE STUDY 2.3   GRANITE, SCHIST AND DISCONTINUITIES, CAMERON HIGHLANDS, MALAYSIA

[4°34.1'N 101°20.0'E]

It is possible to deduce the almost complete geological history of the area from a single road cut (Figure CS2.3.1a), close to the Cameron Highlands. The outcrop is situated within the area discussed in Case Study 9.1. The oldest rocks are dark coloured, mica schists with a pronounced, shallow-dipping foliation, which is wavy and contains lenses of white quartz. Some of the quartz lenses have asymmetric augen and one appears to be the remnant of an isoclinal fold hinge. The contact with the light-coloured granite dips steeply into the cut, but on a micro-scale is highly uneven. Close to the margins of the medium-grained granite, there is a finely layered band that is concordant with the granite contact. Small, irregular bodies of pegmatite occur close to the granite contact and in places are intruded as narrow dykes into the schist. A sketch of the geological components of this outcrop is shown in Figure CS2.3.1b.

GEOLOGICAL HISTORY

The sequential events, from the oldest (1) to the youngest (5) that make up the geological history of this outcrop, provide a template for the region as a whole:

1. Sedimentary deposition of mudstone/siltstone layers.
2. Burial and lithification of the sediment.
3. Deformation, dynamic metamorphism (mica-rich schists) and quartz vein intrusion related to regional tectonic processes.
4. Continued deformation with intense shear of the pre-existing structures – the low inclination of the foliation would suggest a thrust style of faulting.
5. Intrusion of medium-grained granite with flow segregation of the minerals within a distinct band close to and parallel with the intrusive contact. Possible, contact metamorphism of the schist indicated by the dark colour related to the crystallization of a fine-grained amphibole (Fe/Mg mineral).

ENGINEERING CONSIDERATIONS

- Abrupt changes in rock strength of the different lithologies.
- Relatively planar, low-friction, penetrative discontinuities in the schist could, in other localities, adversely dip out of the natural or cut slope.
- The presence of a sub-vertical contact and igneous layering close to the ground surface could weather preferentially, particularly in tropical regions, thereby lowering the soil/rock interface.
- The granite/schist contact is welded and strong.

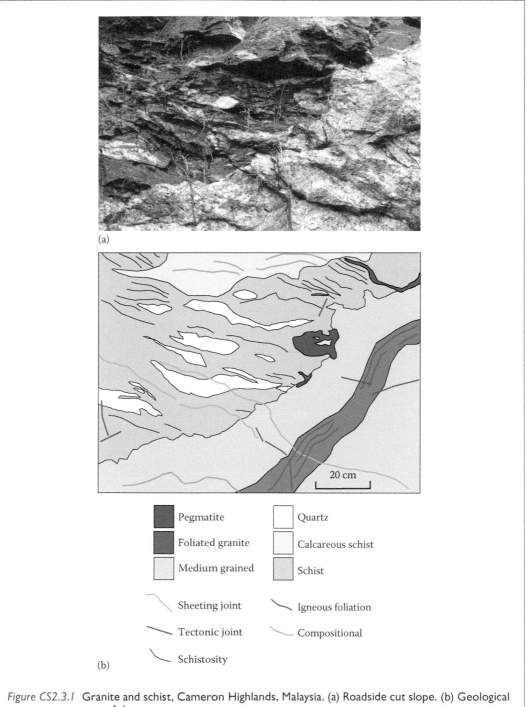

(a)

(b)

Pegmatite

Foliated granite

Medium grained

Quartz

Calcareous schist

Schist

Sheeting joint

Tectonic joint

Schistosity

Igneous foliation

Compositional

20 cm

Figure CS2.3.1 Granite and schist, Cameron Highlands, Malaysia. (a) Roadside cut slope. (b) Geological map of the exposure.

CASE STUDY 2.4   WEATHERED TUFF AND COLLUVIUM, TOLO CHANNEL, HONG KONG

[22°27.40′N 114°17.95′E]

A low sea-worn cliff exposes variably weathered coarse ash crystal tuff overlain by recent colluvium – hill slope deposits (Figure CS2.4.1a). The homogeneous tuff is cut by a moderately dipping shear zone with a weak foliation and is intersected by two sets of joints. A series of the thin quartz veins cut the tuff. Decomposition of the tuff is controlled by the joint pattern leaving moderately decomposed, subangular rock cores surrounded by brownish, iron-stained highly decomposed rims, with light green, completely decomposed tuff along the shear zone and within the relict joints. The decomposition of the tuff above the shear zone is more intense than below it. The tuff is unconformably overlain by colluvium containing variable percentages and sizes of angular fragments of slightly weathered tuff. A thin basal layer of sandy silt with many small angular rock clasts is overlain by a unit of stiff, silty clay with scattered clasts, which contains a thin silty clay layer near the top of the exposure. A large fragmented block of rock lies on top of the weathered tuff. From this outcrop alone, it is impossible to determine the relative ages of the two geological units, and all that can be inferred is that the deformation and weathering of the tuff preceded the deposition of the colluvium. This outcrop also provides an indication of the hydrological regime within the rock/soil mass in that the shear zone acted as the main pathway for groundwater, with water also moving along the joints, particularly above the shear surface. A sketch of the geological components of this outcrop is shown in Figure CS2.4.1b.

GEOLOGICAL HISTORY

The sequential events (from 1, oldest, to 6, youngest) that make up the geological history of this outcrop provide an indication of what may be expected in engineering projects within the region:

1. Deposition, burial and lithification of a relatively homogenous crystal-bearing tuff during a volcanic episode (Jurassic).
2. Deformation of the rock sequence and formation of a moderately dipping shear zone (thrust fault?) with associated shear foliation.
3. Formation of two, roughly orthogonal, joint sets.
4. Uplift, erosion and tropical weathering.
5. Decomposition of the tuff – most intense along joints and the shear zone.
6. Sequential deposition of layered colluvium (Quaternary and recent) derived from fresh rock exposures further up slope during rainstorm (basal layer) and possible landslide events. The clay layer could be an ancient soil horizon.

ENGINEERING CONSIDERATIONS

The following points provide an indication of the geological phenomena that may be significant for an engineering project in the immediate vicinity of the exposure:

- Distribution of weak, decomposed material (soil) and hard rock fragments highly variable.

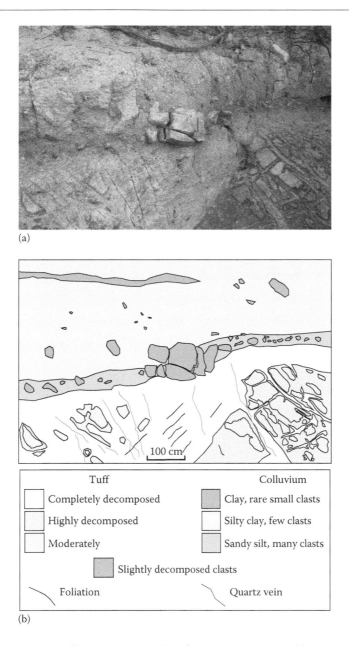

(a)

(b)

Tuff	Colluvium
Completely decomposed	Clay, rare small clasts
Highly decomposed	Silty clay, few clasts
Moderately	Sandy silt, many clasts
Slightly decomposed clasts	
Foliation	Quartz vein

Figure CS2.4.1 Weathered tuff and colluvium, Tolo Channel, Hong Kong. (a) Low sea cliff exposure. (b) Diagram of the geological relationships of the exposure.

- Completely decomposed rock layer dips beneath less decomposed material. See Case Study 11.1 for an example of this situation.
- Hydrogeology of the ground influenced strongly by completely decomposed shear zone within the tuff.
- Layer of colluvium composed of strong blocks of rock in a silty clay matrix.
- Deposition of colluvium during multiple events on an active hill slope.

Chapter 3

Geological maps

Basic geological mapping has been at the core of geology since its inception in the late eighteenth and early nineteenth centuries when it was realized that rocks of similar composition and fossil content could be traced across country. The most famous first geological map was of England, Wales, and part of Scotland by William Smith, a canal engineer, who single-handedly completed the geological map by 1801. It was published as a hand-painted, 8 ft tall by 6 ft wide edition in 1815 (Figure 3.1), and this unique map can be viewed at Burlington House, London, the headquarters of the Geological Society of London. The map colours he used for the different strata were light purple for the oldest rocks that crop out over most of Wales and sea-blue for the narrow ribbon of Carboniferous limestone that snakes its way across the country. Similar colours are still used in modern-day geological maps of the United Kingdom. Most importantly, a geological cross-section of England was included on the map which predicted the strata beneath the countryside, thereby providing the first 3D view of the geology. This map was an amazing achievement and its overall accuracy has borne the test of time – it is well worth reading the fascinating story behind the creation of this map – *The Map That Changed the World* by Simon Winchester (2001). Over the following two centuries, most of the world has been geologically mapped at various scales, but the impetus to teach and continue this essential skill is diminishing, even within the national geological surveys – a great loss.

Geological maps and their accompanying cross-sections present a 3D picture of the distribution of rock and superficial deposits in the ground. Geological information is also provided on derivative maps, including regolith, geological engineering and hydrogeological maps, together with geophysical maps and photogeological maps. All provide valuable information on the composition and structure of the rocks and superficial deposits in the broad context of geological information and need to be studied in the assessment of ground conditions for any engineering project. In this chapter some of the methods used in the production of geological maps are highlighted in order for the engineer to be familiar with the different techniques used in their construction. The reader is referred to excellent and comprehensive textbooks on the subject for full coverage of this fundamental branch of geology (e.g. Bennison, 1985; Boulter, 1989; Maltman, 1998). However, more importantly for the engineer, this chapter focuses on how to interpret geological maps and assess their accuracy and reliability.

The publication of geological maps is dominantly the responsibility of national or regional geological surveys, largely financed by governments. However, geological departments in universities have also produced excellent geological maps as adjuncts to scientific papers or reports in the past, but this is becoming less common now as research moves away from basic geological mapping.

Figure 3.1 Geological map of England and Wales by William Smith published, 1815. (Courtesy of British Geological Survey © NERC 2015. CP15/071. All rights reserved.)

ENGINEERING CONSIDERATIONS

Geological maps are the most important data source for the formulation of any geological or ground model. They include published geological survey maps, figures in scientific journals and attachments to project reports and research theses. Where little or insufficient geological information is available, mapping of the project area may be required. The importance of geological maps and the argument for continued geological surveying and map modification using all available information has been succinctly argued in a paper by Nowell (2014), in which he reiterates the time-honoured assertion by S.R. Wallace: *There is no substitute for the geological map and section – absolutely none. There never was and there will never be. The basic geology still must come first and if it is wrong, everything that follows will probably be wrong.*

Detailed geological mapping of any development site, infrastructure project area or landslide is of paramount importance to any investigation. It is essential that the engineer is fully aware of the local and regional geological framework in order to propose and substantiate any geological and ground model. Engineering ground investigations require accessing and studying all available geological maps, not only of the site itself but also the surrounding areas, and this is especially important in areas of complex ground conditions, or where particular adverse geological features may have more than local significance. Increasingly, more detailed geological assessments of the subsurface geology are required for new engineering projects, particularly offshore or in those areas with few rock exposures. The new data from site investigations and geophysical surveys provide a better understanding of the geology of the area and may necessitate modification of the existing geological maps. Geological maps are not definitive and require updating as more data become available.

To comprehend the geological map sufficiently well for a constructive dialog with the geologist responsible for the modelling of the ground conditions, the engineer must consider the following points:

- Geological maps not just contain information of the distribution of the rock types and superficial deposits, faults and structural data but also provide the geological framework for the project site.
- The scale at which the area was geologically mapped and subsequently published will influence the accuracy of the geological linework.
- The geological cross-sections provided with most geological maps refer to specific transects across the area and must be used with caution when assessing a particular site not on those transects.
- Geological maps can evolve over time as more data become available. The date of the survey is therefore of vital significance, as modifications of the geological linework and the legend may be required using more recent unpublished data, borehole information and geophysical data.

The engineer should be aware that many national geological surveys have continual updating programmes and therefore must be contacted for their most recent interpretation of the geology of an area – this may be in digital format or on unpublished map sheets. Some geological maps or archival field maps also provide information on actual traverses made during the survey, thereby providing an indication of the actual coverage of data collection, which could impinge on the accuracy of the linework.

Many of the chapters and case studies in this book include geological maps and cross-sections to illustrate the distribution of the rock and superficial deposits, geological structures and contact relationships, for example granite and dykes in Hong Kong (Case Study 5.1), volcanic and sedimentary rocks from Scotland (Case Study 6.2) and metamorphic rocks from Nepal (Case Study 7.1).

TYPES OF GEOLOGICAL MAP

The basic framework of the geological map has not changed radically since the time of William Smith; however, the format and composition of the legend has evolved, and there are now a range of derivative maps that focus on specific issues, for example engineering geological and hydrogeological maps. Also the mapping of the superficial deposits (e.g. alluvium, glacial deposits) that cover the bedrock has now become standard.

Maps are produced at a range of scales dependant on the availability of mapping data and the objectives of the survey – whether to show details of the geology in important areas of research and economic development or to give a regional, countrywide perspective. Most countries have geological surveys, which as part of their remit are to publish geological maps and store geological data. In the United Kingdom, for example, the geological map coverage by the British Geological Survey at 1:50,000 scale is almost complete and published maps at 1:10,000 scale are available in the most urban areas, whereas in Hong Kong, a relatively small but intensely built-up region, the Hong Kong Geological Survey, Civil Engineering and Development Department has published a set of fifteen 1:20,000-scale geological maps, which is supplemented in the most developed areas by 1:5,000-scale geological maps. In more remote areas, the only available maps may be 1:250,000 or even 1:1,000,000 scale; however, these only present an overview of the regional geological setting of an area and therefore must be consulted, but it is rare that they will be sufficiently detailed or accurate for site engineering purposes.

Geological maps can cover specific areas at the outcrop level, for example in excavations, dam sites and along tunnels, where the rock and superficial deposit distribution, discontinuity patterns and weathering characteristics are documented and their relationships surveyed. These large-scale maps are of vital importance in any engineering investigation as they provide a readily understandable representation of the ground conditions and geological evolution of the project site and surrounding areas. For example, the road cut map of granite and schist in Malaysia (Figure CS2.3.1b), where the geological relationships are fully exposed, not only provides a documentation of the rock types, discontinuities, spatial contexts and contact relationships but also can act as a template for the production of smaller-scale geological maps in adjacent areas, where the outcrops could be few or inaccessible.

Solid geological maps

Solid geological maps are those that only display the distribution of the rocks as exposed on the ground surface or inferred to outcrop beneath any superficial deposits. In offshore areas, a thick layer of marine superficial deposits generally covers the seabed and an extension of the onshore solid geology into these areas is rarely attempted, except where geophysical surveys such as marine gravity and magnetic surveys have been undertaken (see Case Study 5.1). The mapped rock units are commonly coloured and annotated and refer in the main to their overall composition (e.g. mineralogy, geochemistry, fossil content), texture and grain size, stratigraphic divisions (groups, formations, etc.) or the age of the rock (intrusive complexes). A description of the rock units is given in the legend to the map, which may take several forms (see 'Geological map interpretation' section). Commonly, the rocks may be covered by a thin cover of soil, and it is customary to deem the rocks as outcropping at surface if the soil cover is less than one metre thick; however, it is stressed that on solid geological maps, no account of the weathering intensity of the rocks is considered. Cross-sections that almost always accompany the geological maps provide a 3D view of the solid geology and give additional information as to the evolutionary geological history of an area.

Solid and superficial (drift) maps

These show the distribution of the rocks and the various types of superficial deposits, which include glacial drift, alluvium, marine deposits, laterite and wind-blown sands. Inferred boundaries between the rock units beneath the superficial deposits are drawn using information from boreholes and geophysical surveys and extending the solid linework from adjacent areas.

Two examples of solid and drift editions of geological maps from different geological regimes, one of igneous intrusive rocks and the other of sedimentary rock strata, are presented with comments on the geological relationships seen on the map face and the format of the accompanying legend. The first map is part of the geological map of Yosemite Valley, California, United States (Figure 3.2). The satellite image of this area is shown in Figure 9.2a, which highlights several joint sets of different orientations – a feature important to the

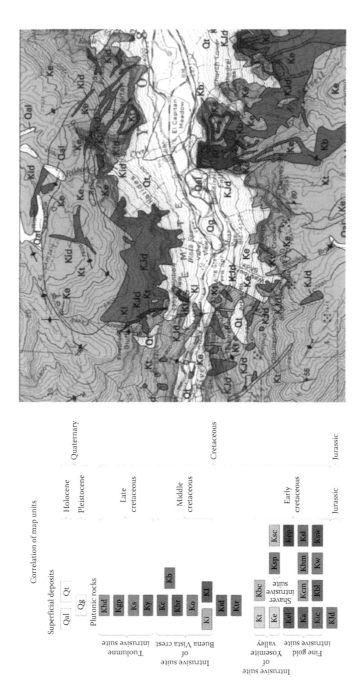

Figure 3.2 Part of the geological map and legend of the Yosemite area, United States. (Peck, 2002. Courtesy of United States Geological Survey.)

engineer but not shown on the geological map. The rocks are composed of a range of plutonic rocks that have been grouped into a series of 'suites' defined by their mineralogy and age. As the exposure in this mountainous region is extremely good, the linework for all contacts between rock types is depicted as unbroken lines. Structural data are confined to a few dip and strike symbols for brittle fault/joint orientations, which for engineering purposes do not provide a statistically representative sample. No solid linework has been extrapolated beneath the superficial deposits that are thickest in the main valley area. Superficial deposits, which are divided into alluvium (Qal), glacial deposits (Qg) and talus (Qt), are concentrated along the main valley floor and at the base of cliffs and steep slopes. The legend to the map shows the different rock types as annotated tablets that are grouped into plutonic rock suites and divided by age of intrusion. The map sheet also provides a summary of the composition of the different rock and superficial deposits as running text. From the map face it is possible to determine the form of the intrusive rock bodies: sub-horizontal sills (Kb) to the south of Yosemite Meadows, a vertical dyke (Kt) at the south-east corner of the map and dome-like upper contacts to a pluton (Kt) located to the north of the main valley. No cross-sections are provided.

The second geological map is from the Mendips area of South West England (Figure 3.3) where a bedded series of Carboniferous Limestone has been mapped around an asymmetric anticline (British Geological Survey, 1962). This is overlain unconformably by Triassic

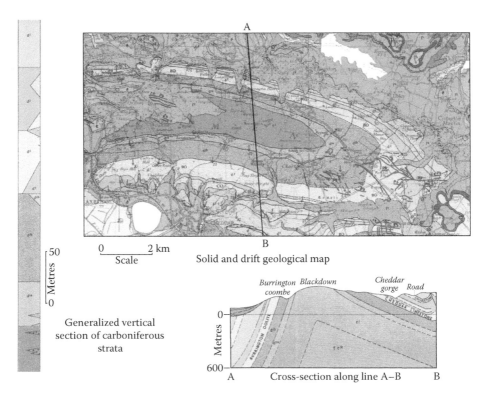

Figure 3.3 Part of the geological map sheet of the Bristol District, South West England. (British Geological Survey, 1962) The cross-section is along the line shown on the geological map. Note that the vertical scale of the cross-section is three times that of the horizontal scale. The generalized vertical section displayed is only part of the complete sequence exposed in the district. Thicknesses of the units may vary across the map sheet. (Courtesy of British Geological Survey © NERC 2015. CP15/071. All rights reserved.)

conglomerates that cascaded from high ground at that time. Most of the linework is continuous implying its position is accurately known, although most of the area is covered by farmland with a relatively limited number of exposures. Here the surveyor would have had to rely on mapping of low topographic features, such as convex and concave breaks in slope, that can be traced across country and rock fragments in the thin overburden. Where the mapped contacts between rock units have had to be inferred by extrapolation or no topographic features could be mapped, there was a degree of uncertainty and the depicted linework is shown as broken. No possible extensions of the solid geology linework beneath the superficial deposits are suggested. Bedding readings are given as dip arrows and no joint orientations are given. The legend is presented as a stratigraphic column with rock units named and their thicknesses drafted to scale. Several cross-sections accompany the map, and these provide a good representation of the subsurface geology to depths of 600 m. The map displays different apparent thicknesses of the rock layers on opposite sides of the main anticline – on the northern limb where dips are between 75° and 50° to the north, the units appear narrower in comparison to those on the southern limb where dips average 25° to the south. Details of the composition of the sedimentary rock units are provided in an accompanying memoir. In more recent map sheets of the British Geological Survey, it is common for the map face to provide more details of the rock units as text and may include geophysical or other maps.

Derivative maps

The geological maps have been put to many uses, and derivative maps have been compiled using the information from these, together with additional information related to a particular aspect or aspects of interest, including engineering, land use planning, water resources, depth to bedrock, mineral and oil exploration, subsidence, geomorphology and landslide susceptibility. Of these, only regolith, engineering geology and hydrogeological maps are described here. However, it must be stressed that although these maps may be in an easily understandable format, some are derived by algorithmic computation of several data sets and therefore may give unsubstantiated results, which are open to misinterpretation. Derivative maps should always be studied in consultation with the original geological map.

The *regolith* is any material exposed at the Earth's surface that is not rock and includes all superficial deposits and decomposed rocks. It may strongly influence the stability, hydrogeology, tunnelling parameters and foundation characteristics of the ground, and therefore it is vitally important for ground engineering investigations. A regolith map displays the distribution of the different types of regolith, for example alluvium, talus, highly decomposed rock and residual soil, with less emphasis on the variations in exposed rock types (Fletcher et al., 2002; Sewell and Fletcher, 2002). These maps are therefore distinct from solid and drift maps, which do not show the distribution or thickness of weathered rock. However, although regolith maps can be a very useful component in some landslide and other engineering investigations, they are not readily available, and most have to be commissioned on a project by project basis.

Engineering geological maps are those that are focussed on particular geological or manmade features that are considered to be important for engineering purposes, for example the delineation of areas with subsurface mining, collapsed solution features (e.g. swallow holes), contaminated ground and unstable slopes. They also include geomorphological data, which identify natural or man-made topographic features such as landslide scarps, made ground, boulder trails and heave mounds.

The factors that influence the flow of water is of paramount importance in many engineering projects, and hydrogeological maps may help in the engineering assessment of a project area. They display the distribution of the rocks and superficial deposits with reference to their permeability, identify the main aquifers and locate any water courses or springs in the area. They are based for the main part on geological maps, which rarely show these features.

MAKING A GEOLOGICAL MAP

Making a geological map is not as simple as it appears, for the geologist has not only to identify the type of rock or superficial deposit but also has to define what is actually mappable and what is not. For instance, a particular thick, white sandstone bed may be recognized, followed and mapped over a large area, but a slightly thinner sandstone bed may be very similar to many other sandstone beds in a sedimentary sequence and therefore not distinct enough to be mapped separately. Every geologist has his or her own particular technique in making a geological map, but overall there is general commonality in the mapping methods and how the data are presented, thereby providing a broad worldwide consistency. It may appear simple to be able to draw contact lines between two rock types in the field, but in reality these contacts are very rarely exposed over any significant distance on the ground surface, except in glaciated and mountainous terrains where there is nearly 100% exposure, along cliffs and in quarries. Actual contacts may be exposed in isolated exposures or along river courses, but they can rarely be located with such precision in adjacent areas. As will be explained in the section on reading a geological map, the confidence in the location of a geological contact is highly variable and not only reflects the degree of rock exposure but also more operational constraints such as accessibility and manoeuvrability to and within the area to be mapped and the time allocated for the mapping project.

The traditional process of making a geological map involved field mapping and meticulously walking over the ground following particular sedimentary units, igneous boundaries or fault traces. However, such methodology is very time consuming, and in places where access is poor, for instance in jungle terrains, it is often better to use satellite or aerial photographs.

Basic principles

The geological map is the intersection between the land surface and the geology. Thus the form of the linework will depend on the topography, for example horizontal and very shallowly dipping strata or igneous contacts will approximately follow the topographic contours (Figure 3.4a), whereas sub-vertical strata will cut right across the terrain (Figure 3.4b). Using intersections between particular geological contacts and the contours, it is possible to construct accurate cross-sections and block diagrams. Figure 3.5 displays the relationship between topography and a moderately dipping bed where stratum contours, lines of equal height on the bed, can be drawn to allow an accurate cross-section to be constructed. For the engineer it is only necessary to be able to recognize such relationships on the geological map as construction of cross-sections is mainly the responsibility and expertize of the geologist. Figure 3.6 is a schematic geological map of bedding – unconformable, intrusive contacts that intersect an undulating topographic ground surface. These geometric relationships can be recognized on published geological maps (see Figures 3.2 and 3.3).

One of the basic techniques of geological mapping is the measurement of planar surface (bedding, foliation, etc.) orientations in the field, which are defined by dip azimuth and angle of dip (Figure 3.7). These data assist in the making of accurate cross-sections and can be used in the statistical analysis to define joint sets and fold geometries and orientations.

(a)   (b)

Figure 3.4 Rock strata/topography intersections. (a) Low dipping sedimentary and volcanic rock strata outcropping subparallel with topographic contours. In places the harder layers form topographic ridges even though the rock is not exposed, Tibet. (b) Steeply dipping limestone unit overlain by thinly bedded limestone and mudstone unit cutting right across topography, Spain.

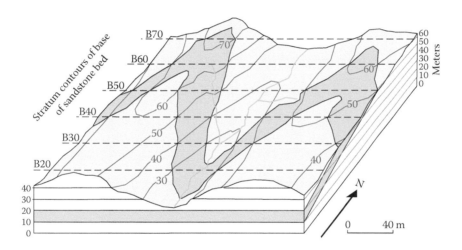

Figure 3.5 Block diagram of a gentle, southward dipping rock sequence. The stratum contours of the base of the sandstone bed (brown) are parallel to the strike of the sequence, and its dip, measured perpendicular to the strike, can be calculated as 20°. The true thickness of the sandstone bed, measured perpendicular to the base of the bed, is 10 m.

Where the geology is poorly constrained, the geologist has to rely on geological extrapolations, geophysical surveys and an intimate knowledge of the geological processes that have actively controlled the distribution and structure of the rock and soil. It is, therefore, essential that the geologist is fully aware of the local and regional geological framework in order to propose and substantiate any geological model. For example, fault patterns mapped on the regional scale or observed on satellite images are commonly replicated on a larger scale and even at the site level or in rock exposures. Increasingly, more detailed geological assessments of the subsurface geology are required, particularly offshore or in those areas with poor rock exposure. It has become essential, therefore, that the understanding of the geology is continually reassessed, geological maps updated and databases maintained.

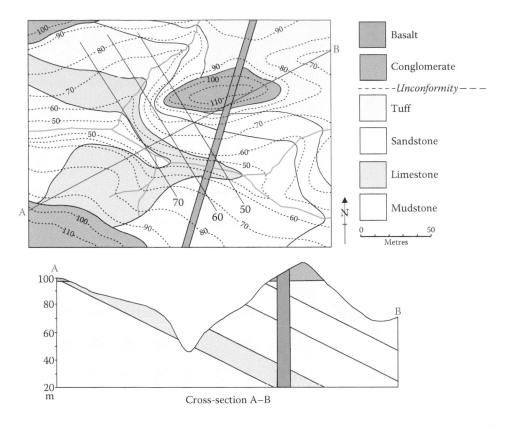

Figure 3.6 Hypothetical geological map displaying the main outcrop patterns for a moderately NE-dipping sedimentary–volcanic sequence, overlain unconformably by a horizontal conglomerate and intruded by a basalt dyke. The cross-section has been constructed along the line A–B using stratum contours – those shown are for the top of the limestone bed.

Figure 3.7 Orientation data for a planar surface. Strike is measured in the direction of a horizontal line on a planar surface, the dip azimuth is at right angles to the strike in the direction of maximum slope, and the angle of dip is the maximum inclination of the planar surface measured from the horizontal. The vertical shaded plane on the picture is perpendicular to the strike.

Satellite images

The use of satellite images over the last few decades has been probably the most significant advance in the making of geological maps. They can provide an accurate regional picture of the geology and in many places can be more accurate than ground surveys and display detail that is not readily apparent in the field. The use of multispectral satellite images adds other dimensions to the understanding of the geology of the area. Some spectral bands will highlight the difference of light reflectivity of different rock types, whereas others will show ranges of moisture content of the ground or vegetation patterns, both of which reflect the underlying geology. In addition, in many areas satellite images depict the continuity of fault traces and fold patterns with distinct clarity, whereas on the ground their exact location and regional extension are not so obvious. The use of satellite images is paramount for the geological mapping of remote areas, for example in many jungle-covered regions of the world, the images provide a detailed representation of the geology, while the lack of rock exposure, vegetation cover and the difficulty in access make conventional field mapping a challenge and less accurate. The geological map derived from the images can be ground checked in critical localities to provide descriptions of the rock types, identification rock sequences, measurement of the geological structures and information on contact relationships.

Examples of satellite images that display the geological contacts are seen in the sedimentary sequences in Colorado, United States (Figure 9.2a); gneissic banding near Aswan, Egypt (Figure 9.2c); and folded strata in the Andes, Peru (Figure 10.2a). The reader is encouraged to view these images for themselves and explore other areas, for example the country around Harrisburg, United States [40°21′N 76°57′W]. There, beds of the resistant rock strata form low topographic ridges that are heavily wooded, and these outline a series of folds that can be traced across the area. Each ridge represents a particular bed within the sedimentary sequence and the intervening low ground is underlain by softer beds. From this image, comprehensive and accurate geological maps could have been drafted without recourse to field work, although spot ground checking would provide a more comprehensive product. Any engineering project that is of a regional nature should regard satellite images as an essential tool in the understanding of the geology and may warrant computer image enhancement and analysis of the different spectral bands to highlight particular geological features.

The use of the Google Earth application, which provides images of the world using a mosaic of satellite and conventional aerial photography, is a valuable resource for the assessment of the geology of an area and should be studied for all ground engineering projects. The images define the regional context of the project site, thereby alerting the engineer to any geological features that might cross the site. Examples of Google Earth images covering a variety of geological environments are illustrated throughout this book and in some cases provide important additional information that is not included in the geological map, for example the different weathering characteristics of two intersecting granite plutons, Huang Shan, China (Figure 5.12a), and distribution of joint sets in a granite, Yosemite, United States (Figure 9.2b).

Aerial photographs

Aerial photographs, particularly when studied using a stereoscope, can provide extremely detailed information on the geology, geomorphology, hydrogeology, subsidence features and site history. However, aerial photograph interpretation (API) requires an experienced professional to understand and document the many natural and man-made features that can be identified on the images, particularly in built-up areas. The procedures and theory behind API are succinctly outlined in Boulter (1989).

Since the 1940s many areas have been surveyed by conventional aerial photography. These photographs provide a valuable resource as the photographs may have been flown in different years, thereby providing a chronology of the evolution of the area in the recent past. Identification of features such as areas of made ground, old excavations, water seepage, unstable slopes, cliff erosion and changes in drainage patterns can be made through studying a sequence of old photographs, whereas this information may be lost or hidden today. Aerial photographs are particularly useful in connection with natural terrain landslide studies (Parry and Ruse, 2002; Geotechnical Engineering Office, 2004).

GEOLOGICAL MAP INTERPRETATION

Probably, the engineer's first introduction to the geology of a project area will be the published geological map. These maps will provide the preliminary basis for the planning of the project, including feasibility and risk assessment, identification of possible problems to be encountered, construction of cross-sections, analysis of any environmental issues and development of a ground investigation programme. However, it must be emphasized that these maps are not site specific, and the assumption that the geology of a particular small engineering project site is exactly as shown on the published geological map must be taken with caution. So it is essential that the engineer is fully able to understand the components of a geological map and its legend with reference to the composition of the rocks and superficial deposits, the significance of the linework (formation boundary, intrusive contact, unconformity, fault, etc.) and the confidence level for the positioning of the linework.

Geological linework

Probably, the most difficult task for a geologist is to draw a line on a geological map for it is rare that actual contacts or fault planes are exposed continually over significant distances. For example, the thin rock strata that traverse a hillslope in Tibet (Figure 3.4a) are continuous but do not outcrop along their complete length. As a consequence, many factors have to be assessed before a final decision is made on what is feasible or desirable to map. If one maps in too much detail and splits the succession into many parts, it is inevitable that many of the rock units will be difficult to follow or recognize over a large area; map too expansively and lump large sections of the sequence into broadly defined units, then the meaning and usefulness of the map may be lost.

On most geological maps, the use of dashed lines to denote uncertainty for the location of geological linework provides an indication of the lack of precision in the mapping, albeit the geologist's best estimate at the time of the survey. However, even a solid line does not necessarily mean that the contact or fault was exposed, for the geologist may have used very well-defined topographic features, vegetation changes or photogeological lineaments to position the linework. It is important for the engineer and geologist to be able to identify and assess those areas on the map where there is ambiguity and doubt and therefore require additional ground investigations or geophysical surveys.

Map legends

The legend provides the key to reading the geological map, as it presents information on the divisions shown on the map face: the names, descriptions and annotations of the stratigraphic units (groups, formations, members), plutonic and metamorphic rock suites and superficial deposits. The relative and absolute ages (when known) of the divisions are graphically displayed (Figure 3.3) or their tablets placed in chronological order (Figure 3.2), and in some legends unconformities and lateral thickness variations are presented. A list of the various symbols is always provided which relate to, among others, measured geological features (e.g. bedding, joints, foliations), metamorphic textures, mine working, made ground, geomorphological features (e.g. landslides, fault scarps, swallow holes, etc.) and boreholes.

Every geological map is unique, and different national geological surveys have their own symbolization and ways of presenting the legend. However, all geological map legends will include most of the following information:

- Brief description of the mapped rock and superficial units, including colour, composition and bedding characteristics, together with their map symbols.
- Identification and description of any defined sedimentary/volcanic groups or plutonic rock suites.
- Graphic representation of the variability of the thicknesses of the mapped units by generalized vertical sections or cross-sections.
- Classification of the rocks and superficial deposits with respect to standard chronological terminology (e.g. eras, periods and epochs).
- A key to the symbols on the map face, including those of bedding, joints, schistosity, axial plane traces of folds, faults and shear zones.
- Description of any metamorphic, hydrothermal and mineralization features.
- Location of any boreholes.

Other information on the map sheet should also include the year(s) of the geological survey and publication date of the map, together with the names of the geologists who mapped the ground and the commissioning organization. Reference should also be made to the accompanying memoirs or explanation leaflets and these must be consulted.

Cross-sections

Of particular relevance to the engineer is the creation of cross-sections that represent the subsurface rock distribution below ground surface of a project area. The accuracy of these cross-sections is paramount in the appraisal of the engineering conditions to be expected during the construction phase of the project. No cross-section will be 100% accurate, but it is the skill of the geological professional to provide the most likely solution that is in accord with all the map and ground investigation data and most importantly is in accord with geological processes and history. Such cross-sections will utilize not only the geological maps available but also the previous ground investigation data in the vicinity and data gained prior to and during the project itself. One of the main uses of cross-section for the engineer is the definition of the geotechnical properties of the different subsurface soil and rock masses, and their distribution for design purposes, and to establish the likelihood of encountering unforeseen ground conditions during the construction phase of the project. Remember when things go wrong it will be these cross-sections which will be scrutinized in detail and will commonly form the basis of any claim.

Cross-sections are some of the most illustrative and informative ways to show the subsurface distribution of the geology of an area (e.g. Figure 3.3). The accuracy of cross-sections is largely dependent on the data collected at ground surface but can be reliant on or enhanced by geophysical surveys using, for example, seismic, gravity, magnetic and ground penetrating radar techniques. The oil industry, in particular, is dependent on seismic reflection results to determine the composition and structure of deep sedimentary basins. However, for most ground engineering projects, cross-sections need only be constructed for relatively shallow depths, mostly between 10 and 200 m. For tunnels, especially those under the sea and deep underground excavations, geophysics provides the only way to establish the initial cross-sections prior to any site investigation boreholes being completed.

Cross-sections use orientation data of solid geological features observed at the surface, for example bedding, igneous contacts, faults and fold axial plane traces, and extrapolate them to the subsurface. The validity and reliability of a cross-section will be dependent on several factors, and therefore it is expedient to understand the rationale behind the construction of the cross-section presented. The main questions that should be asked by the engineer in order to assess cross-sections and consider the need for additional ground investigations are as follows:

- What was the density and completeness of the geological data collected data at surface, and were all relevant data used to construct the cross-section?
- Does the cross-section make geological sense in terms of, for example, the continuity of strata thickness, normal geological processes, the geology of the region and borehole data, taking into account the spacing and orientation of the boreholes?
- What are the accuracy and weak points of the cross-section in terms of the extent of extrapolation from known surface data or between boreholes, how much is conjecture, and what features could have been missed?

Chapter 4

Sedimentary rocks

Sedimentary rocks are either made up of broken and eroded fragments of pre-existing rocks that have been transported to their site of deposition by the action of water, ice, wind or gravity (e.g. sandstone and mudstone) or formed by the accumulation of the remains or secretions of animals and plants (e.g. fossiliferous limestone and coal) or by the precipitation of minerals from solution (salt and other evaporite deposits). These three processes have resulted in the formation of detrital, biological and chemical sedimentary rocks, respectively.

The chemical, physical and biological changes that occur after the deposition or accumulation of the sediment are termed *diagenesis*, which includes the reworking of the sediment by organisms, the loss of water by compaction and the binding of the constituent grains by interstitial cements, together with the complex process of solution and redeposition of material from within the sediment itself, but excluding surficial weathering processes and metamorphism. The overall transformation of a sediment into a rock is referred to as *lithification*. Most sediments contain large amounts of trapped water between their component grains, and during diagenesis this is expelled due to compaction of the sediment by the weight of the overlying strata. The degree of loss of water depends on the original grain size of the sediment and the composition of the grains. For example, the water content of a mud reduces by over 80% during diagenesis to form a mudstone, whereas many sand bodies will retain more voids in the resultant sandstone, unless they are subsequently filled by secondary cements. It is the voids in the sedimentary rocks that largely determine the porosity of the rock and the capability of storing oil and gas and will greatly influence its permeability.

During the evolution of the Earth's crust, the components of sedimentary rocks have often been reworked many times during successive sedimentary cycles which are made up of several phases – sedimentation, burial/compaction, diagenesis/lithification, uplift and erosion (Figure 4.1). For example, in Tibet, the horizontally layered sequence of conglomerate, sandstone and siltstone in the foreground of Figure 4.2 was deposited in a lake that was situated along the northern flank of the rising Himalaya. The detrital components of these rocks were derived from the erosion of sedimentary and other rock types exposed in the Himalaya that, in part, belong to an older sedimentary cycle. Compaction and diagenesis of the lake sediments were followed by uplift of the complete rock sequence related to the continuing upward building of the Himalaya. Currently, this sequence is itself being eroded by a network of incised rivers, which transport the derived materials downstream to be redeposited across nearby alluvial planes or distant deltas. These sediments form the start of the youngest sedimentary cycle that is continuing today.

This chapter describes sedimentary rocks in terms of their composition, architecture, texture and bedding characteristics and relates these to their probable depositional environments. Examples of present-day depositional environments are provided using satellite imagery in order to improve the reader's comprehension of variable character and distribution of ancient sedimentary rock sequences. There are many books devoted to the classification,

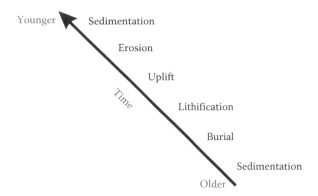

Figure 4.1 Sedimentary cycles. Each sedimentary cycle consists of a series of stages as shown on the diagram. Over geological time, there may be several sedimentary cycles related to different tectonic episodes. Figure 4.2 displays at least three such cycles: the older sedimentary cycles seen within the Himalaya, a cycle of sedimentation related to the deposition of the lake sediments and finally the start of the youngest cycle with the deposition of alluvium in modern rivers.

Figure 4.2 Sedimentary sequences. The flat-lying sedimentary rocks exposed in the foreground were deposited in a lake that once lay along the northern flanks of the rising Himalaya mountain chain, seen in the background. These lake sediments are composed of interbedded sandstone and conglomerate, Western Tibet.

composition and depositional environments of sedimentary rock, of which Tucker (2001), Stow (2005), Collinson et al. (2006) and Boggs (1992) are recommended for further reading and reference. The concepts of sedimentology and stratigraphy are described by Nicols (2009), and the relationships between sea level change and sedimentary facies are detailed in a collection of review papers published by the Geological Association of Canada (1992).

ENGINEERING CONSIDERATIONS

The engineering properties of sedimentary rocks are wide ranging on account of the different strengths of the component grains, composition of the matrix, permeability, weathering characteristics, fracture potential, and the spacing, thickness and continuity of beds within a sedimentary sequence. In addition, rocks that contain appreciable percentages of carbonate minerals, in particular calcite ($CaCO_3$), are susceptible to solution by groundwater, which commonly results in the formation of karst features such as underground caverns and collapse structures (discussed in Chapter 12). Near-surface weathering of limestone-bearing sequences may result in solution depressions and crystallization of secondary minerals (e.g. Case Study 4.1). Areas in which carbonaceous deposits, such as coal and peat, are present need to have special attention due to past underground mining activity and the possibility of the migration of methane. Exploitation or dissolution of evaporite deposits, such as salt, also has led to subsidence at the surface, and in places the evaporite can flow upwards due to overburden pressures. Weak rocks such as chalk may also contain extremely hard nodules of siliceous chert that can adversely affect tunnel boring machines.

Knowledge of potential variations in the distribution, composition and architecture of individual layers or units within a particular sedimentary rock sequence is an important element in the assessment of a project area in which sedimentary rocks are present. This may be achieved by reference to the depositional environments in which the rocks were deposited, thereby providing a template for the geometry of the units that make up a sedimentary rock sequence. Comparison with the type, compositional variations and distribution of the sediments being deposited today provide useful indicators of those features found in ancient rock sequences. As a consequence, modern-day sedimentary environments, some of which may be studied on satellite imagery, are presented in the 'Sedimentary environments' section.

The hydrogeology of sedimentary rocks is extremely variable, ranging from highly permeable non-cemented sandstone (*aquifers*) to almost impermeable mudstone (*aquicludes*). Interbedded rock units with differing permeability need special attention in many engineering assessments, including landslide mitigation, tunnel construction and dam formation.

The key engineering factors that should be considered in any project where sedimentary rocks are present include the following:

- The composition and form of rock units is extremely variable, including planar thick sheets, thin interbeds of different rock types and sinuous ribbons.
- Variable geotechnical properties of the rock strata both vertically and laterally.
- Sharp transitions between rock types with radically different strengths, permeability and fracture potentials.
- The potential for carbonate, evaporite and carbonaceous deposits within the sedimentary sequence needs to be addressed.

SEDIMENTARY ENVIRONMENTS

Knowledge of the different environments in which sedimentary rocks were deposited is a vital tool in fully understanding the 3D form, composition and possible rock associations that may be encountered in any sequence of sedimentary rocks within a project area. For the main environments that will be described in this section, reference has been made to modern equivalents, some of which can be viewed and studied on satellite images. These environments have their ancient counterparts in the geological record stretching back over hundreds of millions of years. However, the location of these environments has migrated over time; as continents moved across the Earth's surface, new oceans were formed, mountain chains uplifted and

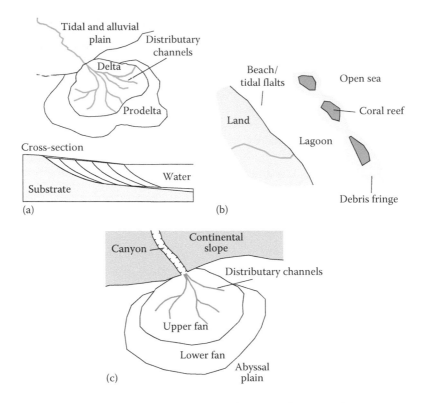

Figure 4.3 Schematic diagrams of the main architectural features of three sedimentary environments: (a) Nearshore, deltaic environment with cross-section illustrating the building out of the delta seawards. (b) Tropical reef environment. (c) Oceanic environment with a turbidite fan system spreading out from the mouth of a submarine canyon.

eroded, sea levels rose and receded and the cover of ice caps fluctuated. The sedimentary processes have left their signatures in the geometry of the rock units, the stratigraphy of the rock sequences, the contained syn-depositional structures and the fossil assemblages. Recognition of these environmental indicators is a key component of any detailed sedimentological study.

Sedimentary environments may be divided into three broad groups: continental environments in which river, lake, glacier and desert sediments were deposited; nearshore and continental shelf environments in which beach, reef, delta, lagoon and shallow water sediments were deposited; and oceanic environments in the deepest parts of the oceans. In the geological record, each of these will encompass a variety of *sedimentary facies*, which are defined by a distinctive combination of rock types and their fossil contents. In this chapter, focus is made on those sub-environments that are common in the geological record. Rocks deposited in glacial environments have been identified, but they are relatively rare and this environment will be discussed in relation to glacial superficial deposits in Chapter 13.

Representative sketches of the distribution of deposits most commonly encountered in the geological record – continental, deltaic, nearshore tropical and oceanic environments – are displayed in Figure 4.3.

Continental environment

The majority of continental sediments were deposited on land as a result of transport by water, gravity, wind or glaciers. Sedimentary rocks that record these erosive and depositional environments are generally not as well represented in the geological record as those deposited in water (see following sections). However, in places, thick sequences of aeolian sandstone have accumulated when desert conditions prevailed over an extended period of time. Examples of continental sediments are more widespread across the present-day land surface where unconsolidated, superficial deposits can cover large tracts of land (see Chapter 13).

River systems carry large volumes of sediment eroded from mountains in suspension as clay, silt and sand and rolled or bounced along the riverbed as boulders and pebbles. This sediment is deposited in river valleys and across flood plains and is termed *alluvium*. The size of the transported material generally becomes smaller downstream away from the mountains, as the energy of the river diminishes and the water is spread over larger areas. Such a system is illustrated in the satellite image of the eastern end of the Himalaya mountain range (Figure 4.4a), where immense volumes of material, eroded from the uplifted

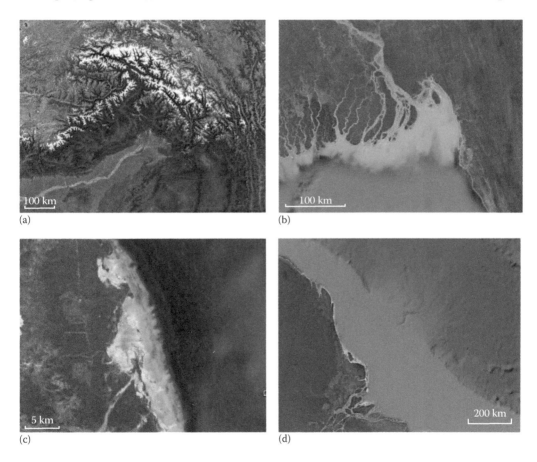

Figure 4.4 Google Earth, Landsat images of present-day sedimentary environments. (a) Continental environment – narrow dendritic valleys filled with coarse-grained sediment passing downstream onto a wide alluvial plain, Eastern Himalaya, Tibet [27°41′N 95°40′E]. (b) Deltaic environment – delta front forming at the mouth of Ganges River, Bangladesh [22°11′N 90°57′E]. (c) Tropical shallow water marine environment with fringing coral reefs, surrounded by carbonate sands passing into finer-grained sediments in deeper offshore waters. Behind the reefs are a variety of depositional environments: lagoons, intertidal flats and beaches. East coast of Andros Island, Bahamas [25°00′N 77°58′W]. (d) Turbidite fan system in an offshore oceanic environment. A submarine feeder canyon is located at the edge of the continental platform and a turbidite fan system with sediment pathway channels are clearly seen on the upper part of the continental slope. Offshore from the mouth of the Amazon River, Brazil [2°37′N 47°45′W].

areas, are first deposited as mainly gravel in narrow incised valleys and further downstream as sand and mud across wide flood plains with restricted river channels being filled with gravel. The sediment is eventually deposited in deltas along the coasts of oceans or continental lakes.

Gravity-transported sediments are those that accumulate along the base of mountain slopes due to materials derived from the weathering of the exposed rocks cascading down the mountain sides or incorporated into landslide debris (see Chapter 13). This is a dynamic environment continually changing in response to uplift rates and climate change, but slower movement also occurs by gradual creep of material downhill. Windblown (aeolian) deposits are predominantly found in two depositional environments: first, in deserts where vast quantities of windblown sand have accumulated in dunes and, second, across extensive tracks of inland areas, where blankets of air-fall dust (loess) have settled.

Nearshore environment

At the interface between the land and the sea (or lake), there are a plethora of influences that control the variations, structures and extent of the sedimentary deposits. These include sea level changes, tides, sediment input from rivers, waves and currents. Not all these can be discussed here, and the reader is referred to the published works for more complete analysis (e.g. Geological Society of Canada, 1992 and references therein). Here, only the three most encountered nearshore environments, in which deltas, carbonate-dominated reefs and beaches were formed, are considered.

A delta forms when a river carrying a sediment load enters the sea or lake and a protuberance of sand and mud is built out from the shoreline. A delta is made up of three depositional environments: a delta plain which is dominated by fluvial processes, brackish conditions and swamps; a delta front that developed beneath sea or lake level where the coarser sediment is deposited; and the prodelta – a deep offshore extension to the delta system where the fine-grained sediment settles out from suspension (Bhattacharya and Walker, 1992). Each of these depositional environments may be clearly seen in the satellite image of the delta region in Bangladesh (Figure 4.4b). The geometry of a complete delta system is schematically shown in Figure 4.3a, which illustrates the extent of separate depositional sub-environments both in plan and in cross-section.

Deltaic rock sequences are composed of alluvial sandstones, restricted to narrow channels, coal derived from organic-rich accumulations in swamps and layers of mudstone deposited during flood conditions (delta plain), stacks of cross-bedded sandstones (delta front) and mudstone-dominated sequences (prodelta). When sea level rises and falls, each environment may be superimposed on each other, thereby forming a stack of successive deltas of different ages.

From the engineering perspective, deltaic rocks can include hard sandstones of different grain sizes either as thick units which may be quarried or sinuous ribbons (ancient river channels) within soft mudstone, coal and carbonaceous shale beds and thick mudstone layers. All these may be repeated several times through the complete sedimentary succession as the sea level rises and falls.

Many limestones were deposited as carbonate reefs and algal mounds on the continental shelf close to land or above submerged islands. Associated with these structures, some which are still preserved *in situ* in rock sequences, there are a variety of reworked carbonate deposits, including fringing sheets of shelly debris and deep water carbonate accumulations transported by turbidity currents. Carbonate-rich sediments may also be deposited on carbonate platforms close to shore in the tidal and beach areas. On intertidal flats, a sequence of algal mats accumulates which are interlayered with lime-rich muds. Commonly, these are associated with the development of salt pans, tidal creek sediments and, in higher-energy sectors, bodies of shell-, pellet- and oolite (small accretionary carbonate spheres)-rich sands. Above sea level aeolian sand deposits accumulate as dunes, which are interspersed with marshy areas. These environments may be identified on a satellite image of coastline along part of Andros Island, Bahamas (Figure 4.4c). A schematic diagram and cross-section of a typical tropical shoreline environment in which calcareous deposits, for example coral reefs, shelly sand and calcareous mud, are deposited is presented in Figure 4.3b.

Limestone provides a particular challenge to the engineer as it readily dissolves leaving voids and underground drainage pathways. Each limestone facies will present different weathering characteristics, and knowledge of their form relative to the enclosing sedimentary rocks is very helpful in assessing a project area.

Oceanic environment

In the deeper parts of the ocean beyond the edge of the continental platform, sediments are deposited either by turbidity currents (or other mass flow mechanisms) down the continental slope or by the thick accumulations of dead micro-organisms and volcanic dust across the abyssal plain (pelagic sediment). Rocks that were deposited by turbidity currents are referred to collectively as *turbidites* that range from regularly thin-bedded sheets of mudstone and sandstone (Figure 4.10b) to debris flows of thick massive sandstone and slumped mudstone (Walker, 1992). The internal structure of sandstones deposited by turbidity currents includes a massive basal part, overlain by sections with graded bedding, convoluted bedding and parallel laminations (Figure 4.11c). Commonly, the sandstone is overlain by mudstone which is in part pelagic in origin. This in turn will be covered by the next turbidite sandstone, whose base commonly preserves casts of structures formed by the erosion of the sea floor by the turbidity current (Figure 4.11d). It is interesting to note that the sandstone bed could have been deposited in a matter of hours or less, whereas the intervening mudstones could have taken thousands of years to have been deposited.

On the continental slope, sediments that have accumulated in shallow waters are transported by turbidity currents down submarine feeder canyons and disgorged as vast submarine fan systems that can extend over a 1000 km from shore. The fan system comprises coarse sediment-filled channels and sandstone-dominated turbidite sequences in the upper part of the fan that passes oceanward to more mudstone-dominated sequences in the lower part of the fan. Many of these features can be distinguished in the Amazon submarine fan of the South Atlantic, as viewed on the satellite image (Figure 4.4d). The turbidity currents are initiated by instability of sediments on the continental shelf, storms, sea level changes and earthquakes. Figure 4.3c displays a schematic representation of the different facies commonly identifiable in a typical turbidite depositional system.

An understanding of the architecture of turbidite systems is important in engineering projects because in some areas the distribution of rock types may be fairly constant, while in others, the relative percentages of sandstone/mudstone units can vary both laterally and vertically. Channel fills of conglomerate and coarse-grained sandstone are narrow and sinuous and may be flanked by relatively weak mudstone.

An important aspect of turbidite fan sequences is that the sandstones can be important oil reservoirs.

DIAGENESIS

Diagenesis is the chemical, biological and physical post-depositional changes in a sediment that is lithified and transformed into a rock. The changes can occur close to the sediment/water interface very shortly after disposition or much later within the sediment pile as compaction, dewatering, cementation and mineralogical changes due to increased pressures and temperatures take place. Diagenesis can have a significant effect on the geotechnical properties of the rock mass, for instance the formation of intergranular cements, hard grounds, nodules and concretions.

Cementation of the grains within a sediment is one of the most important processes that occurs during the formation of a sedimentary rock, as it will largely determine the strength, permeability and weathering characteristics of the rock. The cements precipitate from mineralized water that flows through the sediment after its deposition. The most common cements are the carbonates of calcium, iron and manganese and quartz. Figure 4.5 illustrates two different cements in a sandstone: one, where quartz has grown around the original sub-spherical grains in a desert sand and, second, where calcite has cemented the angular and rounded grains in a deltaic sand. The former is extremely hard, whereas the latter is much weaker and susceptible to dissolution by percolating groundwaters.

During times of minimal sedimentation, the surface of the sea floor can become colonized by many organisms, some of which burrow into the substrate. Precipitation of carbonates within this biozone cements the sediment, thereby forming a *hard ground* above softer sediments, which are in turn overlain by younger sediments as deposition of material recommences (Figure 4.6a). Hard grounds have been found in the Quaternary offshore sequence of Hong Kong (Figure CS4.1.4a) and presented difficulties in their extraction during the formation of the platform for the airport. Hard grounds can also develop in soils and alluvial gravels where strong, impervious iron-rich layers form just below the surface, sometimes referred to as *hardpan*. Present-day beach sands and gravels may be cemented at the surface to form layered beach rock or irregular bodies around iron/wood constructions (Figure 4.6d).

Concretions or nodules form within mainly fine-grained clay-rich or calcareous sediments prior to their lithification. Concretions are sub-spherical to oblate in shape (Figure 4.6c) and composed generally of carbonate, silica and/or sulphide minerals. They often occur

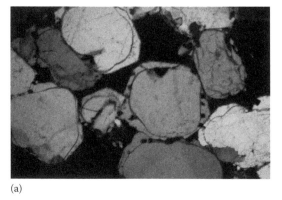

(a)   (b)

Figure 4.5 Cementation in sedimentary rocks. Thin sections of two sandstones displaying post-depositional growth of quartz and calcite as seen in cross-polarized light. View width 4 mm. (a) Quartz overgrowths around sub-spherical quartz grains in aeolian sandstone. The original shape of the quartz grains defined by iron-stained rims. (b) Calcite cementation of angular and sub-spherical quartz grains in sandstone.

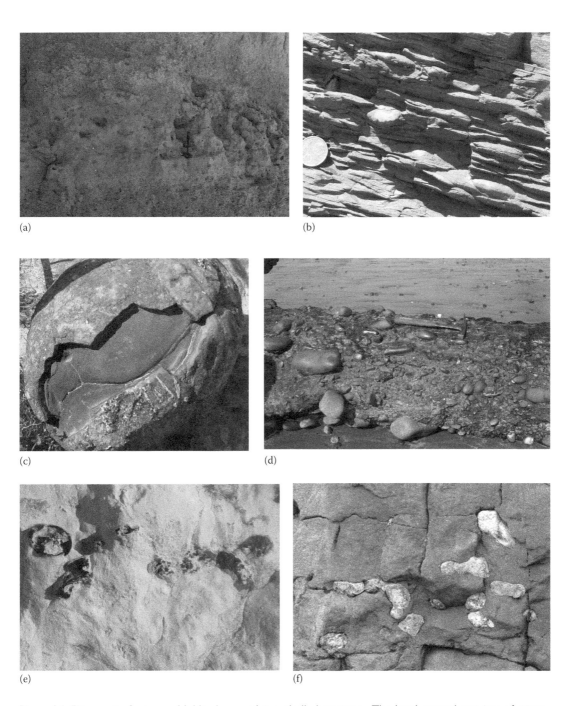

Figure 4.6 Diagenetic features. (a) Hard ground in a shelly limestone. The hard ground consists of micro-crystalline carbonates with relict shells, Algarve, Portugal. (b) Carbonate concretions in shale, northern Spain. (c) Large zoned concretion (30 cm across) with septarian cracks, excavated from sandstone, central Spain. (d) Recent beach rock composed of pebbles and cobbles cemented by iron carbonates, Borth, Wales. (e) Layer of chert nodules in chalk, Eastbourne, England. (f) Irregularly shaped chert nodules in limestone, Anglesey, Wales.

at particular levels within a sedimentary sequence, probably due to specific compositional controls. Concretions commonly nucleate around fossils, and bedding can sometimes be traced through the concretion. Some concretions display radiating cracks filled with calcite. Of particular importance to the engineer are chert nodules in chalk and limestone (Figure 4.6e and f) which are extremely hard and can be problematic for tunnel boring machines in that they are very abrasive or with driven piles if the siliceous nodules form semi-continuous layers.

SEDIMENTARY ROCK CHARACTERISTICS

Sedimentary rocks have extremely varied compositions that depend on a variety of factors, including the sources of the sediment, environment of deposition, chemistry of the water, mineralogy of the intergranular cements and degree of differentiation of the transported minerals by gravity, currents, wind or other mechanisms. However, most sedimentary rocks may be classified in relation to the relative concentrations of the main minerals or mineral groups, namely, quartz, calcite and other carbonates, and clay minerals, which may be presented on a composition triangle (Figure 4.7).

The 3D geometry of most sedimentary rock units can vary from tabular (Figure 4.8a), lenticular (Figure 4.8b), and wedge shaped (Figure 4.8c) to sinuous, all of which are largely dependent on the environment of deposition, although erosion and faulting will in some circumstances modify their shapes. Figure 4.9 provides a generalized portrayal of the common sedimentary rock unit geometries as seen in cross-section. The individual rock units can occur over different scales, for example a single thin tabular unit of mudstone may be traced over hundreds of kilometres, whereas a coarse-grained sandstone or conglomerate may die out in a matter of metres. Knowledge of the depositional environment of a particular rock unit provides essential information as to its distribution, and this can be used when constructing cross-sections from boreholes or extrapolating between outcrops.

The structures that are found within sedimentary rocks provide evidence for the depositional environment in which they were deposited. Environmental factors including whether the sediments were terrestrial, marine or lacustrine, the depth of water, strength of the wind or currents, morphology of the sea or river bed and mechanism of transport of the sediment load can all influence the nature and form of syn-depositional sedimentary structures.

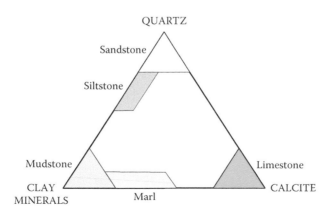

Figure 4.7 Composition of sedimentary rocks in relation to the percentages of the main components.

Figure 4.8 Examples of sedimentary rock geometries. (a) Tabular sandstone beds, Cachopo Valley, Portugal. (b) Lenticular sandstone beds in siltstone. Cut polished rock slab, Glamorgan, Wales. (c) Wedge-shaped pebbly sandstone units in fine-grained sandstone, Algarve, Portugal.

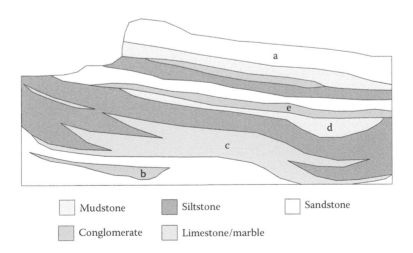

Figure 4.9 Schematic diagram of common geometric shapes and relationships of sedimentary rock units: (a) planar, (b) lenticular, (c) complex, (d) channel and (e) wedge.

Most are small in scale and may be preserved in core samples, thereby providing an indication of the rock associations and geometry of the rock types that might be expected across a wider area. However, more information can be gained where rock outcrops are present and the sedimentary structures can be considered within the context of a complete depositional succession. In this section, the most common and most useful sedimentary structures that can be used in the determination of the original sedimentary environment are described and their significance assessed, thereby allowing a complete and robust geological model to be generated. In particular, they may alert the geologist to the presence of possible adverse geotechnical conditions associated with rocks not present or intersected in the ground investigations or lying at greater depths.

Bedding

Bedding is defined by a change in composition, grain size or even colour and is dependent on many depositional factors such as water depth as controlled by sea level change, current velocity, climate variability and tectonic parameters. It may also define the cessation of sedimentation over a particular period or compositional change due to subaerial exposure or the change in the chemistry of the seawater, for example from oxygenated to anaerobic conditions. Bedding is the most important discontinuity in the majority of sedimentary rock sequences, in that adjacent sedimentary beds can have substantially different compositions, fracture potential, weathering characteristics and engineering properties, for example, an interbedded sequence of strong sandstone and weak mudstone. Bedding is generally sub-horizontal in its original depositional setting, although in certain environments, for example in deltaic or some terrestrial sequences, bedding may be significantly inclined – this is referred to as *cross-bedding*. It may also be reorientated by slumping or post-lithification tectonic movements to become steeply dipping, vertical or even overturned.

Bedding thicknesses are very variable and may change over relatively short vertical and lateral distances. Such variations provide evidence for the environment of deposition and may significantly alter the engineering characteristics of the rock mass. Figure 4.10 illustrates a range of bedding thicknesses and their local variability: a single very thick bedding unit (Figure 4.10a), a sequence of sandstone beds within mudstone displaying different bed thicknesses (Figure 4.10b), limestone beds with fairly constant thickness (Figure 4.10c) and thinly bedded siltstone and mudstone (Figure 4.10d). In places, it is difficult to exactly measure bedding thickness as they may be only defined by imprecise characteristics, for example a train of pebbles within the sandstone sequence (Figure 4.10e). There is a great variability in the sharpness, form and nature of bedding surfaces, due mainly to the syn-depositional processes that were active during the deposition of the sedimentary rock. These are described in the following section. Within any sequence of sedimentary rocks, the bedding may vary from sharp and well defined (Figure 4.10b) to gradational (Figure 4.11b) and diffuse (Figure 4.10e). In addition, bedding surfaces may also be very planar (Figure 4.10c), wavy (Figure 4.11f), irregular in shape (Figure 4.8b) or contorted (Figure 4.11c), in response to currents, sudden changes in water depth, subaqueous erosion or slumping during, or shortly after, the deposition of the sediment. All these bedding features can have a significant effect on the geotechnical characteristics of the rock mass, for example the friction angle of the bedding surface, susceptibility to failure, quarrying methodology, ease of excavation and suitability for building stone. Some bedding surfaces, however, may have no affect on the geotechnical properties of the rock, for example bedding defined by a colour change.

Figure 4.10 Bedding characteristics. (a) Very thick conglomerate bed in a sequence of thinner sandstones, Mount Kailash, Tibet. (b) Sandstone beds displaying variable thicknesses in a steeply dipping mudstone sequence, Aberystwyth, Wales. (c) Limestone beds with comparatively constant bed thickness, Dunraven Bay, Wales. (d) Finely bedded siltstone and mudstone, Peng Chau, Hong Kong. (e) Indistinct bedding defined by trains of quartzite pebbles in sandstone, Central Tibet.

Syn-sedimentary structures

Syn-sedimentary structures provide evidence for the depositional environment of the sedimentary rocks and an indication of the possible composition, distribution and, in some cases, the geometries of the rock units, thereby enhancing the development of the geological model. Sedimentology is a geological discipline in its own right, and this book cannot provide a comprehensive description of all the syn-sedimentary structures that can occur in sedimentary rock sequences. However, a selection of the most common structures is described in the succeeding text, and these should alert the engineer to the importance of the recognition and interpretation of sedimentary structures.

Cross-bedding is an arrangement of thin layers of sediment inclined at an angle to the main bedding planes (Figure 4.11a) and forms mainly in deltaic, windblown and alluvial sandstones. *Graded bedding* occurs in any subaqueous environment in which there is sufficient time to allow a separation (grading) of the component grains with respect to their weight or size (Figure 4.11b). It is characteristic of deltaic, alluvial and turbidite sequences. *Convolute bedding* is a chaotic distribution of the sediment laminae within a distinct sedimentary layer. It is generated by slumping, sliding and load deformation during the deposition of turbidite sandstones (Figure 4.11c). These sandstones may also display sole marking

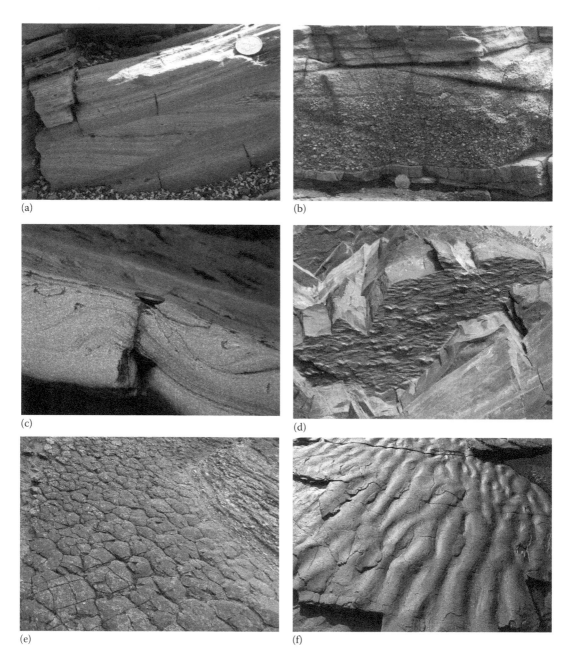

Figure 4.11 Examples of sedimentary structures that provide evidence for the environment of the deposition. (a) Cross-bedding in fine-grained sandstone in a deltaic environment, Peng Chau, Hong Kong. (b) Graded bedding showing an upward gradation from coarse- to medium-grained sandstone in a deltaic environment, Peng Chau, Hong Kong. (c) Convolute bedding in thin sandstone deposited by turbidity currents in a deep sea environment, Aberystwyth, Wales. (d) Sole markings on the base of a sandstone bed deposited by a turbidity current. The long axes of the flute casts indicate the direction of the turbidite flow. Width of view 1.5 m, Aberystwyth, Wales. (e) Mud-cracks in mudstone due to desiccation in a terrestrial environment. Width of view 1.8 m, Anglesey, Wales. (f) Ripple marks on the top of a sandstone bed indicative of wave/current action. Width of view 1 m, West coast, Portugal.

on their bases due to the action of turbidite flow units. The most distinctive markings are *flute casts*, which are knobbly protuberances on the base of the sandstone beds formed by the infilling of erosional depressions in the underlying mudstone (Figure 4.11d). They are commonly elongated parallel to the flow direction of the turbidite. Mud-cracks in fine-grained sedimentary rock are distinctive of a subaerial environment where the sediment has been desiccated during extended periods of exposure to a dry climate (Figure 4.11e). Ripple marks on the tops of sandstones are generated by wave and current action, particularly in shallow waters encountered in deltas and intertidal zones (Figure 4.11f). They are commonly seen on present-day beaches.

CLASTIC SEDIMENTARY ROCKS

Clastic rocks are formed by the transportation and accumulation of broken fragments pre-existing rocks or minerals. Clastic rocks rich in carbonate material (e.g. fossils) or volcanic debris are described in later sections describing limestone and volcaniclastic rocks. Clastic rocks are principally classified according to the size of the constituent grains as conglomerate, sandstone and mudstone. Sandstones are subdivided into fine-, medium-, and coarse-grained varieties, whereas mudstone can also be subdivided into siltstone or claystone dependent upon the proportions of silt and clay fractions (Figure 4.12). Figure 4.13 presents the range of grain sizes observed in core from site investigations in Hong Kong. Although important for classification purposes, criteria other than grain size, such as the composition of the grains (e.g. fossil fragments, quartz and feldspar crystals, volcanic material) or the nature of the cements (e.g. quartz, calcite, iron oxides) that binds these grains, may be more significant for engineering purposes. For example, sandstones with high feldspar or fossil

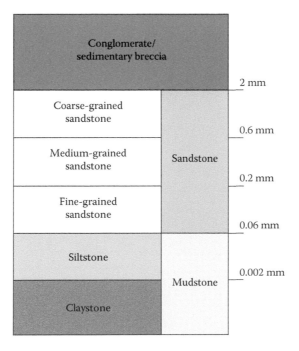

Figure 4.12 Grain-size classification of clastic sedimentary rocks.

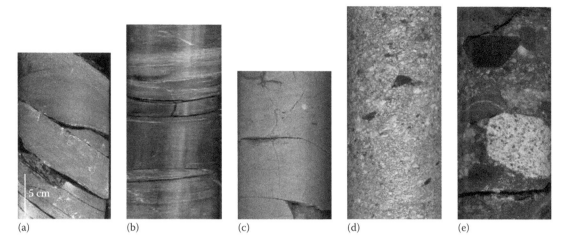

Figure 4.13 Examples of grain size variability in clastic rocks seen in borehole core from site investigations for foundations of bridges and buildings, Hong Kong. Width of core 8 cm. (a) Mudstone. (b) Siltstone/claystone. (c) Fine-grained sandstone. (d) Coarse-grained sandstone. (e) Conglomerate.

contents and those with calcareous cements will have lower strengths than the more quartz-rich varieties, especially when weathered.

Conglomerate

Conglomerates contain large, rounded to sub-rounded boulder cobbles and pebbles (clasts) that were transported by fast rivers, flash floods, or debris flows or accumulated on storm beach deposits. They are indicative of high-energy sedimentary environments in which the originally angular rock fragments have been smoothed by abrasion. The distance from the source of the pebbles and boulders to their site of deposition is relatively short, due to size and weight of the clasts, although in mountainous areas, some extremely large boulders may be transported great distances by rolling along in fast-flowing rivers or flash floods. Conglomerates may display poorly defined bedding, pass laterally into coarse-grained sandstones and be interlayered with finer-grained sediments. They can be found in terrestrial, shallow water and oceanic environments, but they are relatively restricted in their distribution in comparison with the finer-grained sedimentary rocks. The clasts originate from several sources within a catchment area, for example a Quaternary conglomerate from Tibet (Figure 4.14a) contains pebbles of sandstone, granite and volcanic rock, which are indicative of the rock types exposed at the time of the deposition. Similarly, rock core of a much older Triassic conglomerate from a site investigation from Hong Kong (Figure 4.13e) contains the same assemblage of rock types. However, in other cases, the conglomerate may be composed of a single rock type due to the reworking of the pebbles over a long time period and/or the large distances of transport, so that only the hardest rocks have survived the abrasive process. For example, a Palaeozoic conglomerate from Hong Kong (Figure 4.14b) is composed almost entirely of hard quartzite pebbles, which were derived from mountain outcrops that are not exposed today. The matrix of conglomerates is derived from the constituent grains of the boulders, cobbles and pebbles and is generally a medium- to fine-grained sand.

Figure 4.14 Sedimentary conglomerates and breccias. (a) Quaternary conglomerate composed of well-rounded pebbles of granite, volcanic rock and sandstone, Kailash, Central Tibet. (b) Devonian conglomerate composed of mainly quartz pebbles, Bluff Head, Hong Kong. (c) Jurassic sedimentary breccia composed of angular blocks sandstone and mudstone set in a fine-grained sandstone matrix, North Sea oil field, Scotland. Width of core 8 cm. (Courtesy of K. J. Fletcher)

Sedimentary breccia

Sedimentary breccias differ from conglomerates in that the rock fragments are angular and have formed very close to their source, such as in a talus slope below a cliff. The range in fragment compositions in a sedimentary breccia is dependent on the variability of exposed rock types at the time of deposition. For example, variable sandstone and mudstone fragments derived from a Jurassic paleo-fault scarp in the North Sea (Figure 4.14c). Reworking of a sedimentary breccia, for instance by wave action at the base of a cliff, can round some of the fragments and result in minor sorting of the different clast sizes. An excellent example of a sedimentary breccia is exposed along the present-day cliffs at Ogmore, South Wales (Figure 2.3c). There a Triassic breccia accumulated at the base of ancient sea cliffs composed of Carboniferous limestone. The blocks in the breccia are angular, chaotic, highly variable in size and set in a fine-grained calcareous matrix.

It is important to distinguish sedimentary breccias from other types of breccias, for example fault and volcanic breccias, as these have a totally different origins, geotechnical considerations, geological histories and distribution patterns. Sedimentary breccias are generally interbedded with other shallow water and terrestrial deposits and have limited lateral extent. They may have been associated with, and abut against, ancient topographic features, but the recognition or preservation of such features in the geological record is rare. Sedimentary breccias provide important evidence as to the overall geological evolution of an area and can give an indication of the distribution of other rock types.

Sandstone

Sand grains range from rounded to angular in shape dependent upon their mode and duration of transport. For example, windblown sands in deserts have nearly perfect spherical form (Figure 4.5a), whereas beach sand grains are commonly sub-rounded to angular (Figure 4.5b). By far the most common mineral that makes up sandstone is quartz because of its hardness and durability in the sedimentary environment, and when used non-specifically, the term sandstone would indicate that the rock contains above 85% quartz grains. Other minerals include feldspar, mica, zircon and magnetite, and more rarely garnet, apatite and tourmaline. In places, some of these accessory minerals may be concentrated into layers due to their high specific gravity and separation within the water column. The size of the component grains determines the classification into coarse-, medium-, and fine-grained sandstone (see Figure 4.12 for ranges in grain size). Core samples of fine- and coarse-grained sandstone from site investigation boreholes are shown in Figure 4.13. In the fine-grained sandstone core sample (Figure 4.13c), it is difficult to recognize the individual sand grains although a faint bedding is present, whereas angular feldspar and quartz grains are easily identifiable in the coarse-grained sandstone (Figure 4.13d). Sandstones which contain abundant fossil fragments and/or calcareous cement, up to 50% of the rock, are termed calcareous sandstones and are only referred to as limestone when the total carbonate content exceeds this value.

Sandstones were deposited in a wide variety of sedimentary environments from alluvial plains, lakes, beaches, and deltas to offshore areas, where turbidity currents transported sand flows down the continental slope. Each environment produces distinctive sedimentary structures, bedding characteristics and 3D forms to the sandstone bodies. A sedimentary analysis of the sandstones can provide essential data for the construction of a comprehensive geological model of a project area, thereby providing realistic predictions of the variability of the ground conditions to be expected.

Sandstones occur as very thin laminated beds a few millimetres thick to massive sandstones several metres thick, which can be continuous over many kilometres. Other sandstones are lenticular and die out laterally in a few metres or grade into finer-grained sedimentary rocks. The thickness and form of sandstone beds are largely dependent on their depositional environment and therefore an analysis of their sedimentology will provide an indication of the 3D architecture.

Mudstone

Mudstones are sedimentary rocks in which the component grains are less than 0.06 mm across. They are divided into those mudstones rich in silt-sized grains (siltstone) and those in which clay-sized particles predominate (claystone). A siltstone contains mainly quartz with a variable amount of clay and some accessory minerals, such as magnetite. A claystone is composed almost entirely of clay minerals, the most common of which is flaky mineral kaolinite (hydrated aluminium silicate). Examples of siltstone and claystone from site investigation boreholes are presented in Figure 4.13a and b. The colour of mudstone is highly variable dependent largely on the concentration of oil or organic compounds (black), purple/red (ferrous iron) and greenish grey (ferric iron). During compaction and diagenesis, the mudstone may take on a finely laminated structure that imparts a fissility to the rock which is referred to as shale.

Mudstones most commonly occur as thin layers within sequences of sandstone and limestone and other rock types, or as thick almost homogeneous layers, which were deposited either within the deepest parts of the ocean where they may attain a kilometre in thickness or in shallower waters protected from waves or currents. Most of the materials in mudstone settled out of the water column over extended periods of time, in places over many millions of years. Some mudstone layers reflect times of high sea level when water depths were increased across the continental shelf.

Mudstone poses many problems for engineering projects in that it is a weak material that readily converts to mud when disturbed in saturated conditions. Thick layers of mudstone are prone to slow creep and landslides, particularly at times of high rainfall. Mudstone layers are weaker and weather more readily than most other rocks and therefore are preferentially eroded in cliff sections causing collapse of the overlying strata and cliff-line retreat.

Volcaniclastic rocks

Volcanic edifices, ash and lava flows are very prone to weathering and erosion, especially in the marine environment where they can be destroyed over a period of a few years. Quenching of the hot material aids the disintegration of the volcanic rock and releases the more resistive components for redeposition by sedimentary mechanisms such as turbidity currents, wave action or debris flows. The material which survives includes crystal and crystal fragments, pumice, shards, fiamme, volcanic glass and lithic fragments all of which become incorporated into the volcaniclastic sedimentary rock (Figure 4.15). Recognition of the provenance of the volcanic particles in a sedimentary rock may provide essential evidence for the nature of the volcanic eruptions that have taken place in the area, although the actual volcanic centres have been completely destroyed or no longer

Figure 4.15 Volcaniclastic rock composed of fine-grained volcanic material redeposited in the shallow water marine environment. The ripple marks seen in the centre of the photograph were formed by current action and therefore indicate that this sequence is sedimentary in origin and not a primary volcanic deposit, Snowdonia, Wales.

recognizable. The distinction between primary ash flow tuffs and volcaniclast deposits is in places extremely difficult as some of the structures are common to both, for example graded and lamina bedding. Correct analysis of such rocks relies on the recognition of structures that formed when the rock was hot, for example the presence of welded textures, the composition of the adjacent rocks and the architecture of the deposit. Volcaniclastic deposits also include debris thrown into the air by the volcano, including layers of ash several metres thick that can blanket the surrounding areas and bombs of ejected lava and rock fragments that can fall into the volcaniclastic sequences close to the volcano (Figure 6.12).

BIOLOGICAL SEDIMENTARY ROCKS

Biological sedimentary rocks were formed by the accumulation of animal and plant material in the same locality over a considerable period of time, for example coral limestone, algal limestone and coal beds. Such accumulations occur with minimal transport, with the organisms growing on dead ancestors or vegetation accumulating on forest floors.

Carbonate-rich rocks

Carbonate-rich rocks include limestone, dolomite and chalk in which the calcium and magnesium carbonate contents are generally over 90%. They were formed in predominantly tropical seas by either sequential growth in large colonies (coral reefs and algal mats), the deposition of thick layers of microfossils (chalk), the precipitation of calcium carbonate or dolomite cements or the accumulation of shell debris on the foreshore or in shallow seas (bioclastic limestone).

The barrier reefs and atolls grow as isolated masses of calcium carbonate in which the organisms are retained in their original growth positions. As sea level rose or seabed subsided, successive coral colonies grew on a foundation of dead corals. In the geological record, the preservation of intact corals is uncommon due to the delicacy of the organisms and later recrystallization of the carbonate minerals. Figure 4.16a shows a Mesozoic branching coral within a shallow water limestone that is still in its life position. Beach and shallow water accumulations of shells form bioclastic limestone in which depositional sedimentary structures, for example cross-bedding and graded bedding, are very similar to those in other clastic sedimentary deposits. They are extremely varied in their composition and grain size, for example fine-grained comminuted shell debris, intact fossils in a mudstone matrix (Figure 4.16b) and accumulations of large disarticulated shells (Figure 4.16c). A few organisms fall apart on death and form layers of their component parts on the sea floor, for example a crinoidal limestone consists almost entirely of fragmented stem segments and plates that formed the hard parts of the original organism (Figure 4.16d). Some finely laminated limestones were formed as algal mats in the intertidal zone and in places preserve their growth structures (Figure 4.16e). In oceanic environments, the skeletons of planktonic organisms such as foraminifers and coccoliths settle as thick layers over the deep sea floor. On lithification, these layers are transformed into chalk, which is one of the purest limestone deposits (Figure 4.16f).

Post-depositional replacement of the limestone materials by the mineral dolomite (Mg,Ca carbonate) may result in the formation of dolomite or dolomitic limestone. This replacement is due to the percolation magnesium-bearing seawater through the sediment pile and may be patchy or affect the complete succession.

Figure 4.16 Example of different types of calcareous sedimentary rock. (a) Complete coral in its original growth position within a fine-grained Jurassic limestone, Ogmore, South Wales. (b) Complete and fragments of a bivalve and gastropod fossils set in a very fine-grained, calcareous mudstone, Algarve, Portugal. (c) Chaotic assortment of large bivalve shells in a Quaternary bioclastic limestone, Algarve, Portugal. (d) Concentration of crinoid stems within a micritic limestone matrix, Glamorgan, Wales. (e) Finely laminated algal limestone interbedded with lenses of very fine-grained bioclastic material, Aliaga, Turkey. (f) Chalk cliffs with bedding defined by lines of chert nodules, Eastbourne, England.

Carbonaceous rocks

Carbonaceous rocks include coal and oil shale which have formed by the gradual accumulation of organic matter in forest and swamp environments and subsequent lithification and low-grade metamorphism that may be seen in the progression from peat, bituminous coal to anthracite. Coal generally forms in the tropical and subtropical deltaic environment within the inter- and supra-tidal zones where vegetation grew with abundance. Transgression and regression of the shoreline due to sea level changes or tectonic uplift and subsidence has resulted in the formation of stacked sequences of coal, mudstone, sandstone and pedogenic clays. The engineering aspects of underground and open-cast mining and subsidence problems above old coal working are covered in detail in *Coal Geology* by Thomas (2012).

CHEMICAL SEDIMENTARY ROCKS

Evaporation of closed bodies of water, for example a lagoon, inland sea or lake, concentrates the minerals in solution until they crystallize out as a mixture of salts, including gypsum ($CaSO_4 \cdot 2H_2O$), rock salt ($NaCl$) and anhydrite ($CaSO_4$) together with magnesium and potassium salts (Figure 13.9b). The salt flats of Utah and Bolivia are examples of large evaporite bodies that were formed in the recent past and continue to form today. Within rock sequences evaporite deposits can reach over 1000 m thick and represent extended periods of elevated temperatures with no or only periodic rainfall. Evaporites are an important resource and have been mined underground for centuries, which has led to ground subsidence in many areas, for example the salt mines of Cheshire, England.

Salt because of its low density may migrate upwards as domes causing folding and faulting in the overlying strata. These structures are important in the oil industry as salt is impervious to both oil and gas and can become trapped in the folded and faulted sedimentary rock sequences that flank or lie above the dome.

CASE STUDY 4.1   LIMESTONE IN A DESERT ENVIRONMENT, ENERGY INFRASTRUCTURE FACILITY, NORTH-EASTERN JORDAN

[32°33.0′N 39°0.2′E]

C. Roohnavaz, K.A. McInnes and J.H. Thomas
Mott MacDonald Ltd, Croydon, U.K.

A number of onshore energy exploration facilities comprising large heavily loaded reinforced concrete platforms and extensive earthwork structures are constructed on the limestone terrain of north-eastern Jordan desert (Figure CS4.1.1). A major problem with engineering in such remote and often inhospitable desert environment is the scarcity of relevant engineering data and records of natural hazards such as flash floods. The strong limestone bands underlying the proposed facility were interbedded with layers and pockets of porous crystalline limestone/gypcrete. Also identified were a number of 'solution depressions' around the project site. A thorough understanding of the likely geological processes leading to the formation these hazardous features was necessary in order to mitigate the associated risks and engineer a balanced and cost-effective solution for the safe operability of the facility.

Figure CS4.1.1 Relative featureless alluvial plain subjected to flash flooding with surface depressions.

GEOLOGY

The study area comprises Eocene deposits of the Umm Rijam Formation, composed of a sequence of limestone, chert, phosphatic and marly limestone (Abweny, 2009). The superficial deposits overlying the Umm Rijam Formation are described as Pleistocene gravels consisting of unconsolidated subangular chert and limestone clasts and silty sand (Figure CS4.1.2). Site investigation revealed a 200–550 mm layer of predominantly silty clay/clayey silt with fine to medium gravel of limestone, chert and some phosphatic chert overlying the Umm Rijam Formation (Figure CS4.1.3). In some areas, towards the base of this layer inclusion of beige, calcareous

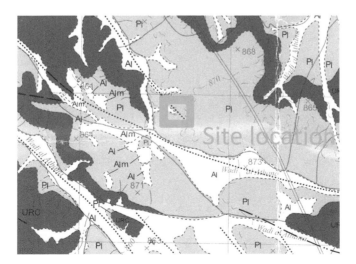

Figure CS4.1.2 Geological map of the site, which is depicted as being covered by Pleistocene (brown) and recent alluvium (yellow) flanked by low outcrops of limestone that form part of the Umm Rijam Formation (red). Grid at 1 km intervals.

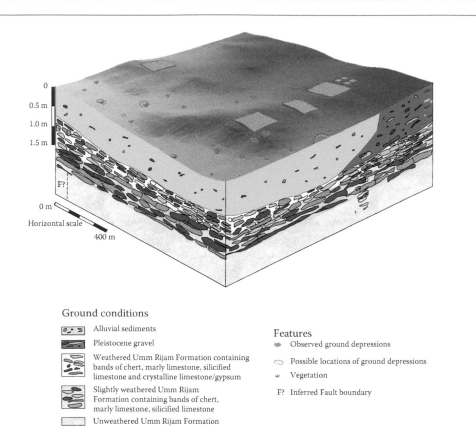

0
0.5 m
1.0 m
1.5 m

F?

0 m
Horizontal scale

400 m

Ground conditions

	Alluvial sediments
	Pleistocene gravel
	Weathered Umm Rijam Formation containing bands of chert, marly limestone, silicified limestone and crystalline limestone/gypsum
	Slightly weathered Umm Rijam Formation containing bands of chert, marly limestone, silicified limestone
	Unweathered Umm Rijam Formation

Features

- Observed ground depressions
- Possible locations of ground depressions
- Vegetation

F? Inferred Fault boundary

Figure CS4.1.3 Block diagram of the near-surface sedimentary and alteration features that have developed in the desert environment in the area of the site.

nodules were identified. The Umm Rijam Formation in the site area comprised a sequence of limestone with bands of chert and 50–350 mm pockets and layers of porous crystalline limestone/gypcrete (Figure CS4.1.4a). The crystalline limestone/gypcrete is orange-brown to white, very weak, friable, porous and fibrous and was located within the top 2.5–3.0 m of the ground (Figure CS4.1.4b). No mention of this porous crystalline limestone/gypcrete was made in the existing geological literature for the area (Abweny, 2009). Mineralogical analysis indicated gypsum as the major component of the crystalline limestone followed by calcite, quartz and trace of palygorsite and hematite. Further analysis confirmed the material to have high calcium oxide and sulphate contents of 35% and 23%, respectively. The chemical analysis indicated 17%–43% calcium carbonate ($CaCO_3$). The high gypsum and calcium contents of crystalline limestone/gypcrete and the ease of dissolution of these materials by percolating groundwater indicate the potential hazard that these present once exposed to water. This can lead to the formation of 'solution depressions' as shown in Figure CS4.1.1. It is likely that 'solution depressions' identified in the vicinity of the project site are caused by the presence of layers and large subsurface pockets of the crystalline limestone/gypcrete.

The formation of the crystalline limestone/gypcrete is not precisely known, but it is understood to have been formed due to the reprecipitation of gypsum from surface infiltration of gypsum-rich water in response to quick evaporation (Brathwaite, 2005). The seasonal conditions

(a)

(b)

Figure CS4.1.4 Near-surface deposits at the project site. (a) A typical section of trial pit exposure with the layered Umm Rijam Formation with interlayers and pockets of porous crystalline limestone/gypcrete. (b) Close-up of gypcrete development within the deposits. Height of sample 15 cm.

in the north-eastern Jordan are semi-arid, experiencing extremely high temperature with heavy intermittent rainfall, which mainly falls during the winter months. The gentle topography of the site combined with clayey/silty soils of low permeability in the top 200–550 mm of the ground lead to ponding of large quantities of rainwater across vast areas around the project site. Dissolution of aeolian-transported gypsum into the rainwater occurs, which then infiltrates the underlying soils and the upper layers of the Umm Rijam Formation. A combination of changes in the ground chemistry and the subsequent evaporation of infiltrated water can lead to the reprecipitation of gypsum within the shallow subsurface, hence defining the crystalline limestone formation to be pedogenic in origin. The formation of gypsum in shallow pedogenic settings is restricted to arid environments where evaporation exceeds precipitation (Eckardt and Spiro, 1999). Fookes et al. (1985) noted that groundwater saturated with calcium sulphate could precipitate gypsum by passing into a carbonate-bearing soil from a siliceous soil. This process appears to have occurred at this site, with gypsum identified beneath chert layers. The depth of formation is limited to the depth of infiltration and the interaction with atmospheric processes (Brathwaite, 2005).

ENGINEERING CONSIDERATIONS

- The presence of crystalline limestone/gypcrete within the region has not been identified in the existing geological literature for the area.
- Flash flooding can occur within the site area.
- Crystalline limestone/gypcrete if exposed to water can be weakened and eventually dissolved leading to solution depressions.

Chapter 5

Intrusive igneous rocks

Intrusive igneous rocks originate from the solidification of magma (molten rock) within the Earth's crust. They include both plutonic rocks, which crystallized slowly in large deep chambers at depths between 5 and 30 km (Figure 5.1a), and near-surface smaller intrusions, which crystallized relatively rapidly in fissures (Figure 5.1b). Intrusive igneous rocks are commonly related by composition and age to volcanic activity at surface; however, it is rare that the actual transition between them is preserved in the geological record. Following the cessation of the igneous activity, the volcanic edifice and the underlying rocks are uplifted and eroded away to expose the underlying plutonic rock and minor intrusions. Many kilometres of rock are worn away during this process, which may take several million years to complete. Figure 5.2 presents a schematic sketch of the relationships between the magma chamber, conduits, dykes, sills and volcanoes and shows a possible level of erosion necessary to expose the deep-seated roots to an igneous complex.

Plutonic rocks can form immense intrusions, called *batholiths*, that are composed of many phases of smaller igneous bodies called *plutons*. Batholiths can outcrop over thousands of square kilometres and can stretch along a complete mountain belt, for example the Trans-Himalaya and Peruvian batholiths. Plutons tend to have sub-circular outcrop patterns and cover areas of only a few tens or hundreds of square kilometres. Although parts of some batholiths may be remarkably homogeneous in chemical composition, crystal content and texture, the individual plutons that make up the batholiths frequently have distinctive compositions and may display zonation patterns of different rock types. Some plutons may have distinct weathering and erosional characteristics due to their composition, joint characteristics and degree of hydrothermal alteration.

Near-surface igneous rocks include sub-vertical *dykes* that were intruded along linear cracks whose orientation was controlled by the regional stress field, and sub-horizontal *sills* that were emplaced as sheet-like bodies between existing rock layers, such as beds in a sedimentary sequence (Figure 10.1). Dykes generally range in width from less than a metre to a few metres, but some can be over a kilometre in width, for example the Great Dyke of Zimbabwe (Figure 5.16). Some dykes may have acted as pathways for the magma to move through the crust and erupt on the Earth's surface as volcanic rocks.

The intersecting Tertiary plutons on the Isle of Skye, Scotland, well illustrate the form of sub-volcanic plutons, the geographic distribution of the largely contemporaneous volcanic rocks and dykes, and the subsequent landscape evolution (Figure 5.3a). There, two plutons of markedly different compositions, one composed of granitic (quartz-rich) rocks and the other of gabbroic (quartz-free) rocks, have been intruded into sequences of sedimentary rocks that are mostly of Precambrian to Jurassic age. The former underlies rounded hills with isolated outcrops, whereas the latter forms steep rock crags along a prominent ridge (Figure 5.3b). Outpourings of lava are preserved mainly to the north of the plutonic centres, and radiating dykes extend for tens of the kilometres.

(a)

(b)

Figure 5.1 Granites and dykes. (a) Outcrop of homogeneous granite body displaying several joint sets that have been preferentially weathered, Ulan Bator, Mongolia. (b) Narrow, dark coloured basalt dykes cross-cutting granite. The dykes have followed pre-existing joints, Po Toi Island, Hong Kong. (Courtesy of the Civil Engineering and Development Department, Government of the Hong Kong Special Administrative Region.)

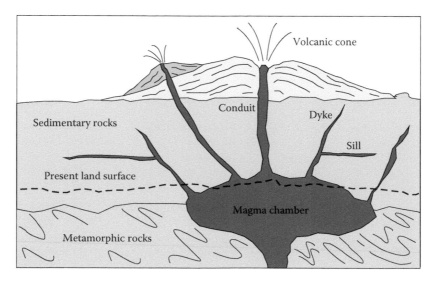

Figure 5.2 Simplified cross-section through an intrusive igneous complex and its associated volcanic edifice. A possible level of present-day erosion is provided.

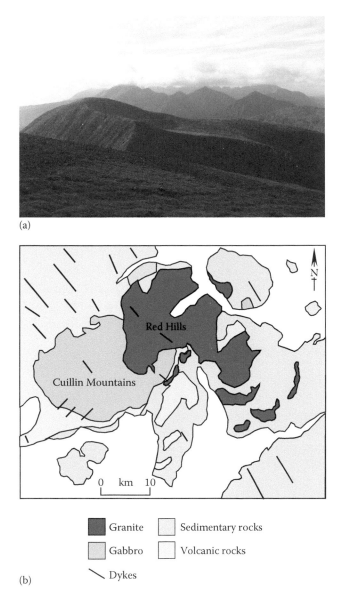

Figure 5.3 Intrusive igneous rock landscape, southeast Isle of Skye, Scotland. (a) View across the rounded granite upland areas of the Red Hills in the foreground towards the jagged gabbro ridges of the Cuillin Mountains on the skyline. Photograph taken from the east. (b) Geological map of intersecting gabbro and granite plutons and their associated lava flows and radiating dyke arrays. (After Stephenson, S. and Merritt, J., *Skye, A Landscape Fashioned by Geology*, Scottish Natural Heritage, 2006, 22pp; Courtesy of British Geological Survey © NERC 2015. CP15/071. All rights reserved.)

ENGINEERING CONSIDERATIONS

The classification of intrusive rocks is at times confusing to the non-geologist as a multitude of rock names has been used to describe these rocks over the centuries. The classification is largely based on the chemical and mineralogical composition of the rock, for example the percentage of the major elements (e.g. Si, Al, Fe, K) or the relative content of the main rock-forming minerals (e.g. quartz, feldspar). These provide an insight into the origin of the primary magma, the plate tectonic setting and the evolution of the magma over time. However, for the most part, these considerations are not of vital interest to the engineer, rather it is more important to understand those features such as contact relationships, joint patterns, alteration intensity and weathering potential that directly affect the engineering parameters, such as strength, permeability and fracture potential. It is these characteristics that have a direct bearing on the ground model, upon which the engineering design is made. However, a basic knowledge of the classification of intrusive igneous rocks and their structures needs to be gained in order to be able to appreciate the construction of the geological model. In this chapter, a simplified classification is presented that covers the most common intrusive igneous rock types that will be encountered in the majority of ground engineering projects.

Plutonic rocks are generally very homogeneous over large rock masses, with the main variations being seen close to the margins of the pluton or across contacts between different generations of pluton (see Case Study 5.1). Quartz-bearing plutonic rocks (e.g. granite, granodiorite) have similar geotechnical characteristics, but local variations between rock types may occur and should be investigated where appropriate. The rarer quartz-free plutonic rock types (e.g. diorite, gabbro, peridotite) have different physical properties to the quartz-bearing varieties in that they have higher plagioclase feldspar contents, specific gravities and magnetic susceptibilities and lower radio-activities, but whether these are reflected in geotechnical characteristics needs to investigated on a case by case basis.

Grain size variations within a pluton are common, particularly close to their margins where more rapid cooling has resulted in fine-grained rocks. However, these contact zones are also associated with very coarse-grained veins and lenses of *pegmatite* that formed as a result of high fluid and gas contents in the upper part of the magma chamber (Figure 5.10e). Such features may affect the spacing and intensity of jointing in these contact zones. Some plutons also have primary flow fabrics defined by the orientation of the larger crystals, and these may impart a plain of weakness to the rock (Figure 5.11a).

The engineer should be aware that it is the post-crystallization features of the intrusive rock that will often be more important in the determination of the geotechnical properties of the rock rather than its actual composition. The most important of these features in engineering terms are jointing, hydrothermal alteration, veining and other contact phenomena, all of which can influence fracture potential, strength, hydrogeology and propensity to weather. Away from the contact zone of plutons, the rocks are more homogeneous and jointing becomes the dominant feature to be assessed. For example, the coarse-grained granite that forms the bulk of a Mongolian pluton has a very constant mineralogy and grain size across a wide area. The dominant and most obvious feature of this granite is the jointing that has controlled the weathering pattern and ultimately the geotechnical parameters (Figure 5.1). The importance of jointing is well illustrated in the rock columns extracted during the formation of a bored pile from Hong Kong. These excavated columns of medium-grained, homogeneous granite display no discontinuities over several metres between narrow bands of weathered, horizontal sheeting joints. Figure 5.4 shows a detail of the granite from columns, which are discussed further in Chapter 11 (Figure 11.13). Similar unjointed sections of granite, up to tens of metres across, are also intersected in deep engineering works.

In summary, the engineer should be cognisant of the following criteria when assessing the ground model of any site in which plutonic rocks are present:

- Fresh plutonic rocks are generally very strong across large rock masses.
- Jointing is commonly the most important feature to be investigated.

Figure 5.4 Detail of a granite columns from a bored pile excavation, Stonecutters Island, Hong Kong. Diameter of column 3 m. The medium-grained granite is homogeneous, and equigranular with no visible joints within the main body of the rock mass. The columns have broken along widely spaced, sub-horizontal weathered sheeting joints.

- Contacts can be associated with hydrothermal alteration, mineralization, veining and possibly faulting, all of which will affect the strength, hydrogeological conditions and degree of weathering.
- Grain size variations near the margins of plutons may influence the spacing and intensity of jointing.
- Primary plutonic rock fabrics may impart a planar weakness to the rock.

Dykes occur as near vertical tabular bodies that cut across the country rocks, and can be traced, in places, for tens of kilometres; as such they commonly define sharp changes in the geotechnical properties of the rock mass (Figure 5.1b). The strength, abrasion characteristics, permeability, intensity of jointing and weathering characteristics, among others, may all change abruptly as one crosses from the country rock into the dyke rock. Dyke rocks have the same range of chemical and mineralogical compositions as plutonic rocks but in general are much finer grained, particularly along their margins where the magma was chilled by the country rock at the time of intrusion. Dykes of different compositions weather in different ways, for example rhyolite (quartz-rich) dykes are more resistant to weathering and form upstanding outcrops, whereas basalt (quartz-free) dykes weather more easily and may form negative topographic features.

The intrusion of different generations of dyke can provide a timeline for the intrusive and tectonic history of an area. In certain circumstances, this can be used to determine the latest tectonic movements in the area – an important factor when assessing the suitability of a site for the construction of sensitive installations, for example a nuclear power station.

In summary, the most important characteristics of dykes for engineering purposes are as follows:

- Dykes are sub-vertical rock masses of different composition and geotechnical properties to the host rock.
- Depending on their composition, dykes can be stronger or weaker than the host rock.
- Joint spacing and orientation may differ between the host rock and the dykes.
- Certain weathered dykes can act as aquicludes due to the high clay content; however, fractured or faulted dyke margins may act as aquifers.
- Dykes may define a major discontinuity in the rock mass and therefore affect its stability and hydrogeological flow regimes.
- Dykes are useful in the determination of the timing of magma intrusion and tectonic activity of an area.

Two case studies of engineering projects are provided as examples of where the form, composition and relative timing of intrusion were essential components of the geological and ground models. The first involved the foundations for one of the supports and roadway spans for a cable-stay bridge in Hong Kong (Case Study 5.1). The bedrock geology consisted of two intersecting granite plutons and a variety of cross-cutting rhyolite dykes, the contacts of which had been displaced and sheared by faulting. The second concerned the construction of a mined station cavern as an extension to the New York subway system (Case Study 5.2). The tunnel was constructed through the upper contact of a granite body, which had intruded a series of highly deformed, mica-bearing schists. A fault zone defined one of the granite/schist contacts and this proved to be a major concern for the engineering works.

COMPOSITION OF INTRUSIVE IGNEOUS ROCKS

Throughout the world, there is a wide variety of plutonic rock types, each with distinctive chemistries and mineralogical make-up, for example granite, granodiorite, diorite and syenite. The simple classification of plutonic rocks, based on the relative percentages of the main rock-forming minerals, is given in Figure 5.5. These percentages will largely determine the properties of the rock, for example geophysical characteristics (e.g. magnetism and specific gravity), abrasion characteristics and weathering potential. Other classifications of plutonic rocks which you may come across depend on the percentage of SiO_2 and other oxides in the whole-rock chemical analysis (*acid, intermediate, basic*) and the relative concentrations of the lighter coloured minerals (*sialic*) against the darker coloured minerals (*mafic*). The quartz-bearing plutonic rocks are by far the most abundant in the Earth's crust, and commonly these rocks are loosely described as 'granite' in many engineering projects. However, it is essential that the rocks are classified more precisely, so that any mineralogical characteristics of each rock type may be documented and also to make sure the geological model as comprehensive and robust as possible. A small selection of the main rock types that you may encounter in the field are presented in Figure 5.6. The reader is referred to treatises and textbooks on the classification, identification and terminology of igneous rocks should the need arise, for example Le Maitre et al. (1989) and Thorpe and Brown (1985).

The identification characteristics of the main minerals that constitute a typical granite are given in Figure 5.5; however, the accessory minerals (<5% of the rock) such as apatite, zircon

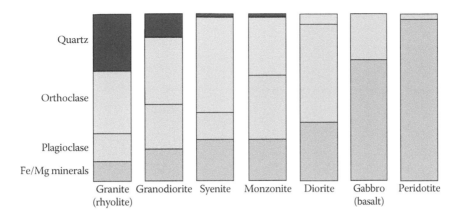

Figure 5.5 Average compositions of the most commonly encountered intrusive plutonic and dyke rocks (in brackets). There are many other rock varieties which are identified by the presence of other minerals, for example some syenites with a low SiO$_2$ concentration contain the mineral nepheline (similar to alkali feldspar but with less silica in its structure) rather than quartz. Fe- and Mg-bearing minerals include olivine, pyroxene, biotite and hornblende. Dyke rock compositions can vary enormously but for engineering purposes may be grouped into rhyolite (pegmatite where coarse grained) and basalt.

and magnetite among others may also be present. Of critical importance in the classification of the quartz-bearing plutonic rocks is the relative percentage of quartz, alkali feldspar and plagioclase feldspar, and this has been used to distinguish between different generations of pluton.

Dyke rocks have much the same chemical and mineralogical composition range as their plutonic equivalents; however, because some dykes are sourced from diverse parts of the magma chamber, they can have rare compositions; some are very rich in mica and amphibole and others may be enriched in rare earth element minerals. However, by far the majority of dykes can be grouped as either rhyolites (quartz-rich) or basalts (quartz-free), which have the same compositions as granite and gabbro, respectively. Some dykes contain large crystals that did not crystallize *in situ* but rather they were transported by the magma from greater depths (Figure 5.17c). In general, dykes are fine grained along their contacts and coarser grained in the central parts due to differential cooling rates of the magma (Figure 5.17b). Some dykes display multiple intrusions of magma with different compositions along the same structural weakness, this results in the formation of composite dykes such those described in Case Study 5.1 and displayed in Figure 5.17d. The engineer should recognize that the naming of all dark coloured dyke rocks as basalt is a great oversimplification, but for ground modelling purposes, it is a good initial approximation.

It is not necessary for the engineer to be a mineralogist, but it is expedient to have a basic knowledge of the identification characteristics of the main minerals that are used in the classification of the igneous rocks. Readers are referred to the many books on mineralogy for a fuller insight into the properties and crystallography of rock-forming minerals (e.g. Hall, 1996), but an excellent introduction is provided in Blythe and de Freitas (1984).

The principal minerals of a granite, as typified by a core sample (Figure 5.7) from a site investigation in Hong Kong, are

Quartz: Light grey, very hard, translucent, no cleavage
Alkali feldspar: Pink, hard, two cleavages, commonly tabular crystals, twinned
Plagioclase feldspar: White or yellowish (especially when altered), two cleavages
Biotite: Dark glossy brown, tabular crystals, very fissile
Hornblende: Dark green, two cleavages, rhomb-shaped crystals

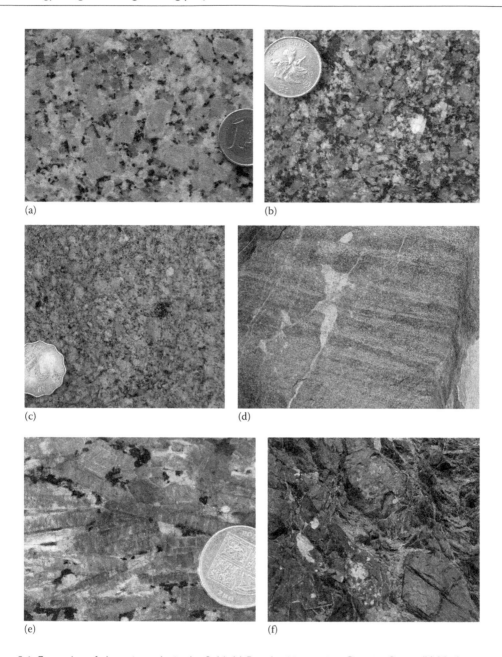

Figure 5.6 Examples of plutonic rocks in the field. (a) Porphyritic granite, Coruña, Spain. (b) Medium-grained granodiorite (rich in Fe/Mg minerals), Cape D'Aguilar, Hong Kong. (c) Fine-grained syenite, Cape D'Aguilar, Hong Kong. (d) Thinly layered gabbro and peridotite, central Tibet. (e) Coarse-grained nepheline syenite with parallel alignment of feldspar crystals, Monchique, Portugal. (f) Peridotite affected by post-crystallization hydrothermal alteration, central Tibet.

The identification of main minerals in hand specimens of intrusive igneous rocks, both in outcrop and in core samples, provides a straightforward and instant, but preliminary, naming of the rock type. However, in many cases, particularly when the grain size is small or the alteration is intense, it is impossible to determine the exact mineralogy. In these cases, it is necessary to have thin sections made of the rock to determine not only the

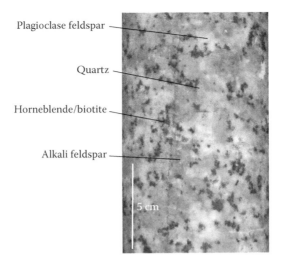

Plagioclase feldspar

Quartz

Horneblende/biotite

Alkali feldspar

5 cm

Figure 5.7 Minerals in a medium-grained granite, core sample, Hong Kong. The characteristics of the minerals indicated are provided in the main text.

details of the mineralogy but also the intergrain textures and alteration characteristics. The contrasting mineralogies and textures of a granite, gabbro and syenite as seen in thin section are presented in Figure 5.8. The intergrain relationships between the constituent minerals of plutonic rocks are of particular importance, for example in the granite thin section shown in Figure 5.8a, the boundaries of the quartz and feldspar crystals are commonly highly indented and embayed. It is this intricate bonding between the mineral grains that provides the high strengths of many granites. X-ray diffraction techniques have to be used to identify the finest-grained minerals that are below the resolution of the microscope or where secondary minerals with indeterminate optical properties are present. This technique is essential in the recognition and analysis of clay minerals that could affect the geotechnical properties of the rock. The engineer must be aware that the results of these additional studies allow the geologist to fully understand the origin and make-up of the rock and therefore provide important information for the creation of a comprehensive ground model.

TEXTURE AND FABRIC OF INTRUSIVE IGNEOUS ROCKS

The terms that describe texture and igneous fabric are common to both plutonic and dyke rocks. *Igneous texture* describes the smaller-scale features of the rock in terms of the size, shape and arrangement of the constituent grains, whereas *igneous fabric* describes the patterns formed by the various shapes and orientations of the component grains. Although the description of the texture and fabric of igneous rocks could be considered to be mainly used for geological modelling purposes, some features are important for engineering projects. For example, grain size commonly influences the joint spacing and weathering intensity, and layering or preferred orientation of the minerals can impart a planar weakness to the rock.

The most common textural term of intrusive igneous rocks is grain size, and this has been used in places to distinguish the different varieties of granite. The standard grain size terms

(a)

(b)

(c)

Figure 5.8 Thin sections of plutonic rocks in crossed polarized light. (a) Granite, Ontario, Canada. K – alkali feldspar with wavy plagioclase intergrowths; Q – quartz; Bi – biotite; Hb – hornblende with intersecting cleavage. Note the serrated edges in the tabular alkali feldspar crystals and the embayed margins in the quartz crystals. View width 5 mm. (Courtesy of P. LeCouter.) (b) Gabbro, Scotland. Pl – plagioclase feldspar with striped light and dark grey crystal twins; O – olivine with vivid interference colours; Py – pyroxene with grey and brown interference colours. View width 5 mm. (Courtesy of the British Geological Survey.) (c) Syenite, Bolivia. K – unorientated alkali feldspar laths; Py – pyroxene with bright green interference colours. View width 5 mm.

used for granitic intrusive rocks are shown in Figure 5.9, but this may be equally applied to all other intrusive rock types. Examples of the different grain sizes of granite from rock cores from site investigation boreholes from Hong Kong are displayed in Figure 5.10. The grain size of intrusive igneous rocks is mainly controlled by the rate of cooling of the magma. Close to the margins of a magma chamber or dyke, where the cooling and crystallization were comparatively rapid, the size of the grains is relatively small, whereas nearer to the centre of the chamber or dyke, where cooling and crystallization were slower, the size of the grains is larger. However, exceptions to this concept do occur. For example, the increased volatile content of molten rock close to the top of magma chambers can promote the crystallization of pegmatite lenses and veins (Figure 5.10e). The relative sizes of the minerals within an igneous rock can be very variable, and this can provide evidence for the

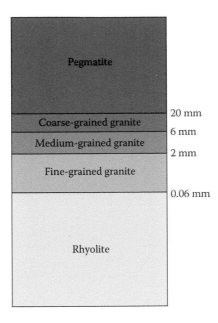

Figure 5.9 Grain size classification of intrusive igneous rocks.

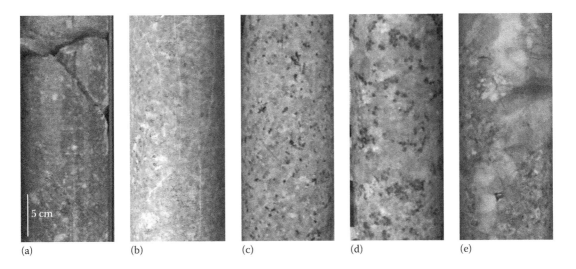

Figure 5.10 Grain size variations seen in granite cores from site investigations in Hong Kong. (a) Rhyolite. (b) Fine-grained granite. (c) Medium-grained granite. (d) Coarse-grained granite. (e) Pegmatite lens within coarse-grained granite.

crystallization history of the rock. The most common terms are *equigranular* (Figure 5.10b), *inequigranular* (Figure 5.6a) and *porphyritic* (Figure 5.11a).

Igneous fabrics include primary igneous foliation, defined by the preferred orientation of mineral grains (Figures 5.6e and 5.11a) and compositional layering of the constituent minerals (Figures 5.6d and 5.11b). The preferred orientation of minerals, such as feldspar

(a)   (b)

Figure 5.11 Fabric and texture of intrusive igneous rocks. (a) Alignment of tabular feldspar crystals in granodiorite pluton defines an igneous fabric, Gredos Mountains, Spain. (b) Primary igneous texture in a gabbro, due to variations in grain size and mineralogy between layers, Skye, Scotland.

tablets, is due to current flow within and close to the margins of magma chambers or along dykes. Compositional layering is formed by variations in the mineral percentages in the rock, mainly due to the cyclical settling out of the heavier minerals, such as olivine, from the magma and their accumulation on the floor of the chamber, or to the selective crystallization of mineral layers of different composition along the sides of the chamber.

PLUTONIC ROCKS

The very slow cooling of the magma within discrete chambers allowed the crystallization of different minerals to take place in a fixed sequence controlled by phase equilibrium criteria. Large crystals grew, settled out and at times were transported within the magma chamber by currents or settled on the floor of the chamber. As a result of these processes, the composition of the remaining magma evolved over time and rocks of different compositions were generated, which were then intruded into the country rock or extruded as volcanic rocks. As more crystals separated from the magma, their boundaries became interlocked thereby providing an incipient strength to the solidified rock mass. Although the classification of plutonic rocks into a plethora of rock names is essential for geological modelling purposes and to provide evidence for the origins of different magma types, the features that are important for engineering purposes are common to most plutonic rock types.

Form

The form of plutons and batholiths, as exposed at ground surface, varies from circular to elliptical for the smaller intrusions to irregular elongate masses for the larger ones. Many plutonic rock bodies display several phases of intrusion, and each of these can in certain circumstances be identified by particular compositional or textural characteristics. An example of this is shown by the two intersecting Mesozoic granite plutons in eastern

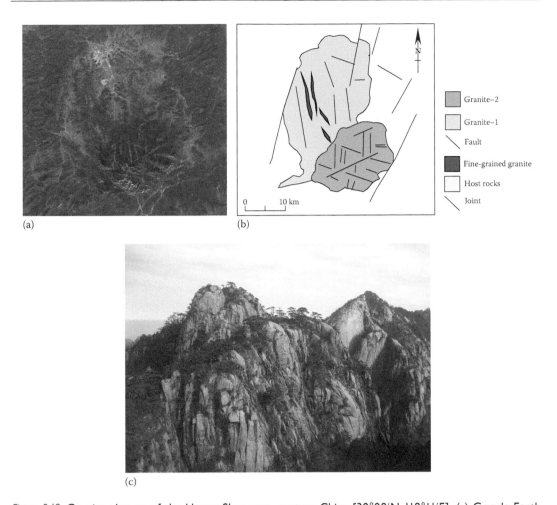

Figure 5.12 Granite plutons of the Huang Shan area, eastern China [30°08′N 118°11′E]. (a) Google Earth, Landsat image of the area illustrates the different topographic expressions of two granite plutons. (b) Geological map of the area showing the outcrop of the two intersecting Mesozoic plutons (1, older granite; 2, younger granite) that have intruded Palaeozoic and Precambrian meta-sedimentary host rocks. Map constructed from satellite image with reference to regional geological maps. (c) Highly jointed homogeneous granite forms the spectacular landscape of the southernmost and youngest of the plutons.

China, which were intruded into a series of old metamorphosed sediments. The different shapes and jointing characteristics of the two granite plutons can readily be seen on a satellite image (Figure 5.12a), and these are represented on the geological map (Figure 5.12b). The oldest elliptical granite pluton is characterized by a subdued topography, whereas the younger, more circular pluton is upstanding, relatively unweathered and displays well-developed, more closely spaced joint sets (Figure 5.12c). The distinction between the two granite plutons, due to a difference in composition, grain size, jointing and/or hydrothermal alteration, has resulted in more pronounced weathering and erosion of the older pluton. However, on the published geological map, they are both described as granite but with different ages. Another example of intersecting plutons, each with distinctive topographic characteristics can be seen in the mountains in the southern part of the Isle of Skye, Scotland. There, the plutons belong to two distinct intrusive phases: an early, upstanding, gabbroic phase and later, topographically more subdued, granitic phase (Figure 5.3).

Contacts

A schematic diagram of a typical upper contact of a pluton, which has intruded a volcanic sequence, is shown in Figure 5.13, and an actual example of such a contact is exposed in a recent quarry face in Hong Kong (Figure 5.14a). The contacts of plutonic rocks with the surrounding rocks vary from relatively simple, with the formation of a fine-grained (chilled) phase along the margins of the pluton, to highly complex margins with the presence of *xenoliths* (exotic blocks of country rock) within the intrusive igneous rock assimilation, hydrothermal alteration and partial melting of the country rock (Figure 5.14c) and *in situ* fragmentation and veining of the country rock (Figure 5.14d). The geometry of the intrusive contacts ranges from sub-horizontal or shallowly dipping across the top of the pluton to steeply dipping on the sides of the pluton. However, quite abrupt changes in the orientation of the contacts may be observed, for example when large, detached, joint-bounded blocks of country rock have split off from the roof of the magma chamber or narrow, dyke-like granite bodies protrude upwards from the main body of the pluton, for example see Case Study 5.1 (Figure CS5.1.4). Further away from the contact with the country rock, the xenoliths become smaller and more rounded, due to partial melting of the margins of the blocks by the molten granite (Figure 5.14b). The country rock is affected by the intrusion of a pluton in three ways: first, by the intrusion of irregular veins of granite, quartz and pegmatite; second, by the emission of fluids and gases that hydrothermally alter the host rock; and finally, by assimilation and contact metamorphism (Figure 5.14c).

The engineer must be able to identify these contact phenomena in rock core so that constructed cross-sections provide a realistic interpretation of the ground conditions to be expected. Although rock cores from site investigations only sample very narrow, generally vertical, sections through the rock mass, it is still possible to identify chilled marginal

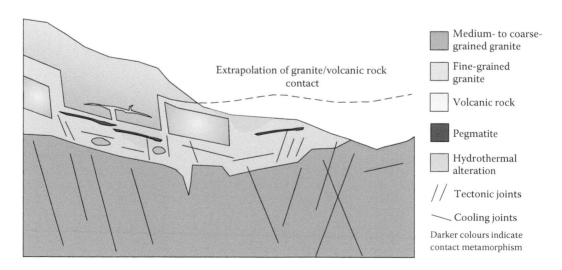

Extrapolation of granite/volcanic rock contact

Medium- to coarse-grained granite

Fine-grained granite

Volcanic rock

Pegmatite

Hydrothermal alteration

Tectonic joints

Cooling joints

Darker colours indicate contact metamorphism

Figure 5.13 Schematic cross-section through a typical upper contact of a granite pluton.

(a)

(b)

(c)

(d)

Figure 5.14 Contact relationships around plutonic igneous bodies. (a) Upper contact of granite pluton with volcanic rocks. Note detached volcanic rock slab above the main granite contact (exposed in the lower part of quarry face) and intrusion of fine-grained granite vein in the volcanic rocks parallel to the contact (exposed in the upper part of quarry face), Anderson Road Quarry, Hong Kong. (b) Rounded partially resorbed xenolith of basalt within porphyritic granite, Cheung Chau, Hong Kong. (c) Assimilation and contact metamorphism of sedimentary rocks close to contact with granite contact, Mount Kailash, Tibet. (d) Angular blocks of sedimentary rock enveloped by granite, Gredos Mountains, Central Spain.

phases (Figure 5.15a), contact relationships and veining (Figure 5.15b), thereby providing valuable information as to the anticipated geology in the surrounding areas.

DYKE ROCKS

Dykes are narrow, igneous intrusions that have penetrated older rocks along pre-existing weakness, for example joints (Figures 5.1b and 5.16) or faults, during a period of tectonic extension (pulling apart). They form in the higher parts of the crust and can act as feeders for volcanic centres (Figure 5.2). Contacts of dykes are normally sharp and follow a general strike direction of joints and faults, which have been determined by the regional

(a)   (b)

Figure 5.15 Contact relationships observed in rock core close to boundaries of plutons. Core samples were recovered from vertical site investigation boreholes, Hong Kong. (a) Contact of medium-grained granite that constitutes the central part of the pluton with a finer-grained, chilled marginal phase of the same granite. (b) Intrusive contact between a younger fine-grained granite with older grano-diorite. Veins of the fine-grained granite have been intruded upwards into the granodiorite pluton.

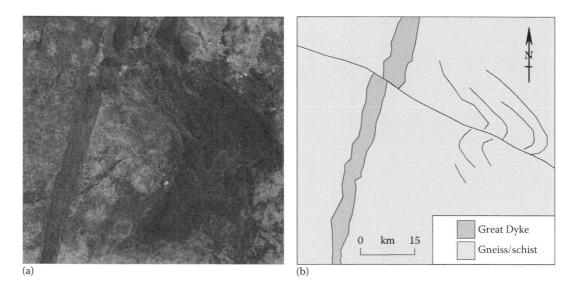

(a)   (b)

Figure 5.16 The Great Dyke of Zimbabwe. (a) Google Earth, Landsat image of the Great Dyke cross-cutting folded Precambrian gneiss and schist [20°27′S 29°44′E]. (b) Sketch of the features displayed in the adjacent image. The dyke has been offset by a right lateral strike-slip fault.

stress pattern. As a consequence, dykes commonly form *dyke swarms* that radiate from a volcanic centre or have similar strikes. Dykes can be very significant in the ground model as they generally have different composition, strength, weathering propensity and spacing of the joints in comparison with the country rocks and therefore may affect the fracturing, hydrogeology, stability and ease of excavation or quarrying of the rock mass.

Form

Most dykes are steeply dipping and vary from a few centimetres to tens of metres in width but in a few places attain widths of several kilometres, for example the Great Dyke of Zimbabwe (Figure 5.16). In places, they may be traced linearly over tens or hundreds of metres, but some extend to over a hundred kilometres. Dykes follow the easiest pathway for the magma to follow and therefore can abruptly change orientation, die out or split into separate 'finger-like' intrusions (Figures 5.1b and 5.17a). Multiple intrusions along the same weakness can increase the width of a dyke (Figure 5.17d). Dykes can radiate out from the parent volcanic centres (Figure 5.3), for example the igneous complexes of Scotland have generated dyke swarms that extend hundreds of kilometres from their source. They can also

(a)

(b)

(c)

(d)

Figure 5.17 Dyke features from Hong Kong. (a) Basalt dyke intruded into rhyolite along two joint sets, Cape d'Aguilar. (b) Chilled margin to a basalt dyke. Note that some of the joints post-date the intrusion of the dyke, Cape d'Aguilar. (c) Porphyritic rhyolite dyke with irregular and indented margins cutting granite. The large feldspar crystals are concentrated in the central part of the dyke, Discovery Bay. (d) Margin of a composite dyke in granite composed of a marginal basalt phase and a later central porphyritic rhyolite phase, Lantau Island.

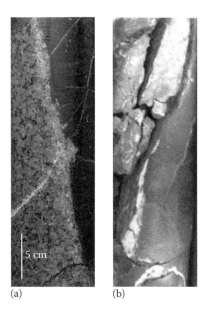

(a)   (b)

Figure 5.18 Dyke contacts seen in vertical boreholes from Hong Kong. Width of cores 8 cm. (a) Fused, micro-indented contact between host granite and basalt dyke. (b) Faulted and veined contact along the margin of a basalt dyke.

occur as ring structures related to the intrusion and collapse of central part of the plutonic bodies or volcanic complexes.

Contacts

At outcrop, many dyke margins are sharp and planar (Figure 5.17b and d) as they follow a particular joint set across an area, but in places, partial melting of the country rock has made the contacts more irregular and indented (Figure 5.17d). The nature of dyke contacts can be of critical importance in the ground model: some contacts may be micro-embayed along crystal boundaries of the host rock (Figure 5.18a), so that the contact the two rock types has become fused and strong, whereas others display post-emplacement faulting, vein emplacement and hydrothermal alteration that renders the contact weak and fractured (Figure 5.18b), thereby imparting a significant discontinuity to the rock mass that could act as an aquifer.

CASE STUDY 5.1   INTRUSIONS AND FAULTS: BRIDGE FOUNDATIONS, STONECUTTERS ISLAND, HONG KONG

[22°19.6′N 114°7.1′E]

The cable-stay bridge connecting Stonecutters and Tsing Yi Islands (Figure CS5.1.1) was completed in 2005. The eastern foundations for the main support tower and the four back spans were sited on reclamation, and therefore the geology was not known in detail prior to the ground investigations. Extrapolations from the known onshore geology suggested that the site was underlain by different phases of granite intrusion, dykes and volcanic rocks, and that one of the most prominent NE-trending faults of Kong Hong could transect the site (Figure CS5.1.2). Inclined boreholes were used extensively to intersect vertical and steeply dipping faults and intrusive contacts. A marine magnetic survey was also conducted over the adjacent offshore areas (Figure CS5.1.3). This indicated that the extension of a major NE-trending fault, mapped onshore and defined by low magnetic values offshore, would certainly pass beneath the site and that E-W-trending rhyolite dykes, exposed on Tzing Yi Island and defined by high magnetic values offshore, would extend as far as the eastern support tower.

GEOLOGY

Two granite plutons with different grain sizes, mineralogies and structural histories are juxtaposed by faults at the site – the Sha Tin and Kowloon Granites (Figures CS5.1.4 and 5.1.5a and b). Both plutons contain large isolated xenoliths of tuffaceous volcanic rock, suggesting that the site is close to the upper margins of the granite intrusions. The oldest Sha Tin Granite to the west is intruded by E-W-trending composite dykes with rhyolite margins (Figure CS5.1.5c) and granite cores. Thus, there are three types of granite at the site, which had to be distinguished in order to establish a coherent geological model and a reliable cross-section. The steeply dipping faults contain gouge (Figure CS5.1.5e) and breccias with the adjacent rocks being sheared and hydrothermally altered (Figure CS5.1.5d), thereby widening the zone of low strength material. The most intense

Figure CS5.1.1 Eastern half of the Stonecutters Bridge and roadway spans nearing completion.

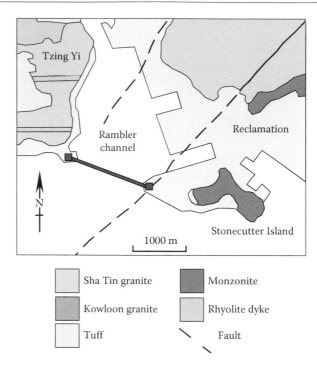

Figure CS5.1.2 Simplified onshore geology of the areas surrounding Stonecutters Bridge faults. (After Geological Map of Hong Kong, Millennium Edition 2000. Published with permission of the Director of the Civil Engineering and Development Department, Government of the Hong Kong Special Administrative Region).

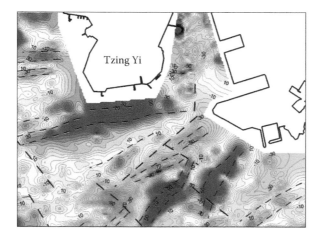

Figure CS5.1.3 Offshore marine magnetic survey in the area around Tzing Yi and Stonecutters Islands. (After Fletcher, C.J.N. et al., 2000b.) Blue, green and red anomalies indicate low, intermediate and high magnetic values, respectively. The NE-trending high values probably reflect a monzonite dyke. (Published with permission of the Director of the Civil Engineering and Development Department, Government of the Hong Kong Special Administrative Region)

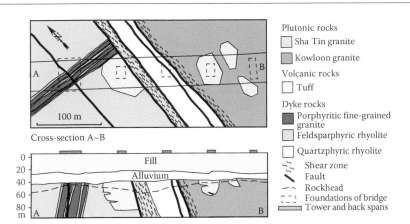

Figure CS5.1.4 Solid geology and superficial deposits on the eastern side of the Stonecutters Bridge. The cross-section A–B follows the alignment of the roadway. (After Arup, 2001.)

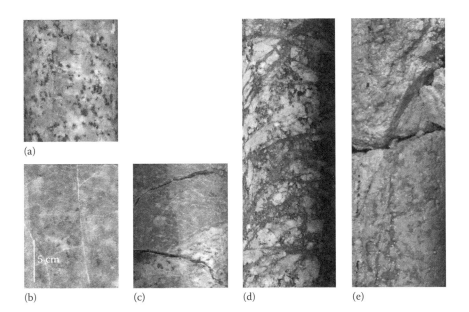

Figure CS5.1.5 Cores from inclined boreholes. (a) Sha Tin Granite. (b) Kowloon Granite. (c) Fused contact between sheared granite and rhyolite margin to dyke. (d) Chloritized, sheared granite adjacent to fault. (e) Sharp contact between weak fault gouge and hydrothermally altered Sha Tin Granite.

shearing was related to a zone of faulting, some 80 m wide, that was considered to be part of a major fault zone. However, here it has been intruded by a rhyolite dyke, which has substantially strengthened the fault zone. Although the site is geologically complex, the ground model is relatively straightforward as the various rock types have similar geotechnical properties, the depth of weathering is fairly constant, and narrow sheared and altered bands did not greatly impact on the foundation design. A cross-section along the alignment based purely on rock type, such as grouping all the granites together, would have been highly misleading and geologically unsound.

ENGINEERING CONSIDERATIONS

- A full understanding of the complex geological history of the site was required to construct the cross-section.
- Inclined boreholes were an essential part of the ground investigation as they intersected the sub-vertical dyke contacts, altered rock zones and faults.
- Fault zones are associated with brecciated, sheared rock and hydrothermally altered zones; however, the impact of the main fault was minimized by the intrusion of a rhyolite dyke.

CASE STUDY 5.2   GRANITE INTRUSION IN SCHIST: SUBWAY TUNNEL EXTENSION, NEW YORK, UNITED STATES

[40°45.37′N 74°00.10′W]

Seth Pollak, *Arup, New York*
Chris Snee, *SneeGeoconsult, New York*

A 2.4 km extension (7 lines) of an existing metro line consisting of TBM-bored single-track running tunnels and a 21 m span × 365 m long mined station cavern (Figure CS5.2.1) underneath midtown Manhattan. Particular challenges include low rock cover to cavern span ratio (0.7), faulted ground and proximity to active rail lines and historic buildings.

Figure CS5.2.1 Station cavern top heading drifts showing typical excavation profiles for granite intrusion (left) versus mica schist (right). The increased overbreak in the schist is apparent.

GEOLOGY

Two major rock types are present along the cavern alignment (Figure CS5.2.2). A central mica-deficient granitic rock, which ranges from pegmatite to medium- and fine-grained granite, has been intruded into medium-grade schists ('Manhattan' Schist) of variable composition and contained some pegmatite. The age of the schists is considered to be Late Cambrian to Early Ordovician, and the granite has a Silurian intrusive age. The upper contact of the granite displays a broad depression across the central part of the cavern section (Figure CS5.2.2). The contact between the granitic rock and the schist is generally intact to moderately weathered, with some notable exceptions. The southern limb of the intrusion is located at the cavern end, where the granite sharply abuts mica schist. The northern limb is characterized by a faulted contact between the granitic rock and quartz–garnet–mica schist (Figure CS5.2.3). The fault zone is approximately 1 m thick and contains decomposed rock and breccia in a matrix of green, low plasticity clay. Adjacent to the contact, the schist is faulted and sheared with the

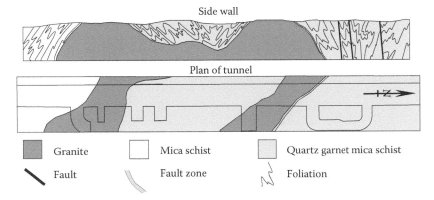

Side wall

Plan of tunnel

| | Granite | | Mica schist | | Quartz garnet mica schist |
| Fault | | Fault zone | | Foliation | |

Figure CS5.2.2 Geological model along station cavern used in the design, showing sidewall (above) and roof (below).

Figure CS5.2.3 Fault zone between pegmatitic phase of the granite and the overlying blocky schist.

(a)

(b)

Figure CS5.2.4 Contact relationships and foliation characteristics. (a) Light-coloured granite intrusive into folded schist exposed in the sidewall of excavation near the surface. The contact to the left is irregular but sharp and follows the foliation in the schist, whereas to the right the contact has been faulted and sheared. A low dipping, weathered, sheeting joint cuts across the lower part of the face. (b) Foliated mica schist where rock dowels were used to 'pin' the steeply dipping discontinuities and prevent planar failure. Sidewall of the cavern access shaft.

development of sub-vertical to vertical foliation fractures and seams that define a series of sub-parallel *en echelon* structures that strike obliquely to the excavation trend. These structures are discrete and bounded by higher-quality schist, similar to that found in the southern end of the cavern. The total length of cavern that was excavated through this poor-quality zone was 100 m. The contact features intersected in the excavation are similar to those exposed on adjacent surface outcrops (Figure CS5.2.4a).

The schist can be observed in two different forms. In the first, there is a lack of jointing and the foliation dominates the rock mass behaviour (Figure CS5.2.4b), and in the second, multiple joint sets can be mapped and the rock takes on a more 'blocky' look. Here the dominant joints have formed parallel to the foliation (Figure CS5.2.2), and the rock mass behaviour is influenced more by the strength of these joints than the rock itself. The granite is typified by two joints sets: an orthogonal set and a sub-vertical joint set. Sub-horizontal joints (sheeting joints) are typically open and clay infilled (up to 15 mm), and moderately to highly continuous with measured trace lengths of 10–15 m (Figure CS5.2.4a). They cut across both rock types and produce the majority of water inflow into the cavern with up to 10 l/min.

ENGINEERING CONSIDERATIONS

- Special excavation measures must be taken in urban areas (controlled blasting, multiple drifts).
- Variable rock types can present differences in ground behaviour for tunnels. These differences must be characterized and assessed by appropriate methods during project design.
- Geological contacts should be defined and thoroughly investigated as they often represent zones of increased alteration, fracturing or groundwater inflow.
- Design-build contract arrangement allowed the designer to have a presence on site during construction, carrying out mapping to validate design parameters such as discontinuity length, spacing and roughness which are often difficult to quantify from boreholes. This resulted in an on time and under budget project delivery.

Chapter 6

Extrusive igneous rocks

Extrusive igneous rocks are erupted from volcanoes, which are the sites where molten rock (magma) from deep within the Earth's interior is erupted through conduits and vents onto the land or sea floor. There are about 500 active volcanoes today of which only about 25 are active in any year. For example, Mount Etna in Sicily (Figures 6.1 and 6.5b) has recently been active with the outpourings of lava and has a long documented history of eruptions over the last thousand years. Their distribution is controlled by major weaknesses in the crust (Figure 2.1) that developed as a result of the movements of tectonic plates (see Chapter 2) and are most common along plate boundaries. However, some volcanoes also occur within continental plates across the world, for example the Mount Suswa (Figure 6.5a) in the East African Rift Valley, and in oceans initially on the sea floor but with time forming new islands, for example the Hawaiian island chain in the Pacific Ocean (Figure 6.3). Eruptions may be violent and short lived, for example Mount Mayon in the Philippines (Figure 6.2), whose most violent eruptions occurred over a few months, or be relatively non-violent and long lived, for example Mount Kilauea (Figure 6.3). Volcanic deposits are commonly inter-bedded with sedimentary rocks, which reflect times of low or no eruptive activity within a volcanic cycle, when sediments derived from the volcanic edifice accumulated on the sea or lake floor.

Volcanic rocks may be grouped into two broad categories: *lavas* extruded slowly at the Earth's surface or on the sea floor and *pyroclastic deposits*, ejected violently from explosive volcanoes as ash-flow and air-fall tuffs. Although many volcanoes are dominantly one category or the other, it is not uncommon for both to be erupted within the life of the same volcano. A schematic diagram of the typical architecture of volcano and the types of deposit is presented in Figure 6.4.

ENGINEERING CONSIDERATIONS

The present-day and recent volcanic activity obviously has significant effect on the planning of engineering projects. Exact predictions of when, how and where volcanic eruptions will take place in the future are still not completely reliable, although imminent eruptions have been foreseen using real-time observations, including the change in the shape of the volcano and increased frequency of shallow Earth tremors. Volcanic hazard maps have been published for many of the present-day volcanoes, and these provide the basis for geological risk assessments of established or future development and infrastructure projects. For example, the location and extent of lava flows that are predicted to erupt from Mount Etna over the next 50 years provide engineers with essential information for planning purposes. The satellite image of Mount Etna shows the location of some of the most recent lava flows in relation to the location of towns and agricultural activity (Figure 6.5b).

Figure 6.1 Mount Etna on the east coast of Sicily, Italy (see satellite image Figure 6.5b). Classic cone shape with the eruptive material, including both lava and pyroclastic flows. The volcano is currently active.

Figure 6.2 Explosive eruption of Mayon Volcano, Philippines, in 1968. Pyroclastic flows cascade down the flanks of the volcano, and the eruptive cloud of ash above the volcano is carried by winds to fall as ash over large areas. (Photograph courtesy of the U.S Department of the Interior, U.S. Geological Survey.)

The edifices of volcanoes that were active in the past are generally not preserved in the geological record due to uplift and rapid erosion. However, the present-day surface of erosion may reveal the conduit of the volcano and the associated intrusive plutonic rocks and feeder dykes within both the remaining volcanic deposits and the underlying basement rocks. In this case, the original summit of the volcano would have been several kilometres above the present day land surface. The distribution and contact relationships of volcanic deposits around recent volcanoes provide a template for understanding volcanic rocks in ancient sequences. It is imperative, therefore, to understand the architecture of the different types of volcano in order to develop reliable ground models to be used in engineering projects. As a consequence, the important features seen in volcanic rocks need to be recognized

Figure 6.3 Vent on the slopes of Kilauea volcano, Hawaii, USA. This active vent has erupted lava through a breach in the summit crater. On the skyline is the silhouette of the low-profiled Mauna Loa volcano. (Photograph courtesy of the U.S Department of the Interior, U.S. Geological Survey.)

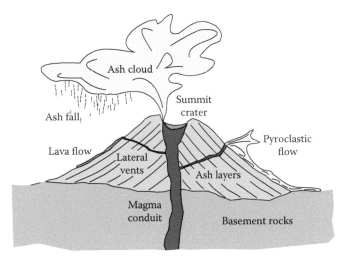

Figure 6.4 Schematic diagram showing the architecture of a volcano and the distribution of the main intrusive and eruptive materials.

and understood so that the probable lateral and vertical variability within the rock mass can be properly assessed in any project area. The engineer should be aware that volcanic deposits can be very weak and easily weathered, whereas others can be extremely strong and resistant to weathering – the two variants can be interlayered or juxtaposed or even grade laterally into one another. Primary discontinuities within volcanic rocks may be important locally, for example columnar jointing and flow foliations in lava flows and pyroclastic deposits. It is the geological modeller that needs to provide the best and most reliable ground model and identify those volcanic features that may impinge on the ground model.

In geologically ancient volcanic rocks, which have possibly been uplifted, faulted and folded, the key considerations for any engineering project are the primary composition, strength, distribution and spacing of discontinuities and 3D distribution of the various volcanic rock units (Figure 6.7), together with the post-emplacement hydrothermal alteration and weathering of the rock units. Many volcanic rocks alter to clays, and these can radically

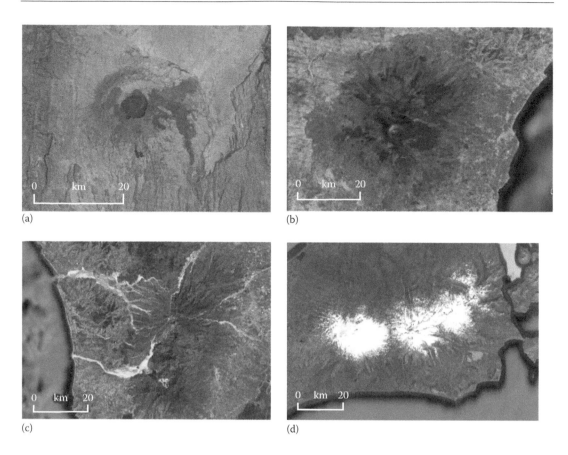

Figure 6.5 Google Earth, Landsat images of volcanoes in different plate tectonic settings. All scales 20 km. The volcano forms are best studied using the oblique view tool on the Google Earth software. (a) Suswa volcano, East African Rift Valley, Kenya [1°9′S 36°21′E]. (b) Mount Etna, Africa–Eurasia plate boundary, Italy [37°45′N 15°0′E]. (c) Mount Pinatubo, Eurasia–Philippine plate boundary, Philippines [15°8′N 120°21′E]. Note pathways of the lahars from the flanks of the volcano to the sea. (d) Shishaldin and Isanotski volcanoes, Aleutian volcanic arc [54°45′N 163°56′W].

affect the physical properties of the rock mass, including its strength and friction angle, and strongly influence the hydrogeology of the site, especially where the clays form a continuous layer. Of additional concern is the presence of swelling clays within weathered volcanic sequences, as these can radically alter the geotechnical properties of the rock and soil mass over time when exposed to the atmosphere.

Many of the features described in this chapter may be seen as unrelated to any engineering problem, but it is emphasized that recognition of these features may provide vital clues as to the most probable distribution of the rock units that make up the ground model and allow predictions of ground conditions. However, it should be realized that joint spacing and their continuity and orientation may have a greater influence on the ground model than differences in the rock type.

In summary, the engineer should take the following into consideration when assessing any site in which volcanic rocks are present:

- Original distribution of the volcanic rocks and awareness to the style of eruption.
- The presence, orientation and nature of any primary volcanic foliations and joints.

- Hydrothermal alteration of volcanic rocks is common and may alter the geotechnical properties of the fresh rock.
- Weathering characteristics of the different volcanic rock types is very varied.
- Clay minerals within weathered volcanic rock sequences can have a marked effect on the stability and hydrogeology of the site.

Two case studies are presented in this chapter in which volcanic rocks were an important element of the geological and ground models. The first is a fatal landslide in Hong Kong (Case Study 6.1), which occurred in highly and completely weathered pyroclastic rocks. The landslide was located in a depression of the rockhead surface above a zone of vertically orientated primary flow fabrics in the tuff. A seam of kaolinitic clay with low friction angles provided the basal slip surface. The second case study describes the geology for the new Queensferry Crossing over the River Forth, Scotland (Case Study 6.2). Volcaniclastic deposits (tuff) were encountered on the northern shore, and these exhibited variations of compressive strength in relation to changes of moisture content following the time since sampling. Associated with these volcaniclastic deposits, an intrusive volcanic vent was intersected at depth that cross-cut the older sedimentary rock sequences.

GEOMETRY OF VOLCANIC DEPOSITS

The study of satellite images of present-day volcanoes provides an excellent way to understand the distribution patterns of the various volcanic materials and to appreciate how quickly erosion, weathering and plant growth can affect these materials today and in the past. In Figure 6.5, the satellite images of three active or recently active volcanoes are displayed – Mount Suswa volcano in Kenya (Figure 6.5a), Mount Etna volcano in Sicily (Figure 6.5b) and Mount Pinatubo volcano in the Philippines (Figure 6.5c). It is suggested that readers view and familiarize themselves with the geometry of these volcanoes as seen on satellite images using Google Earth.

Mount Suswa is an active volcano within the East African Rift Valley, and successive generations of lava that have flowed out of the summit crater can readily be distinguished. These have followed specific topographic valleys on the flanks of the volcano and do not extend a great distance from the base of the volcano. Several centres of eruption can be seen within the summit caldera, which confines the greatest accumulation of lava in the area. The relatively dry climate of the region has minimized the weathering and vegetation growth on the lava flows so that the youngest lava flows maintain their dark colour.

The volcanic cone of Mount Etna (Figure 6.1) has evolved over many millennia as lava has flowed down the slopes in all directions. The Mediterranean climate has allowed rapid soil development and vegetation growth on the newly erupted lava flows so that the sequence of lava flows is, in places, difficult to recognize. The rich nature of the volcanic soils has been exploited since man first arrived in the area so that the agricultural modification of the landscape and establishment of small settlements and towns are concentrated around the volcano. The volcanic hazard to these developments has been substantial, and lava has encroached into the towns in the past.

The explosive eruption of Mount Pinatubo in the Philippines in 1991 was spectacular and of relatively short duration. The top of the volcano was blown off, ash flows cascaded from the summit crater and thick deposits of ash settled over the surrounding areas and even reached Singapore some two thousand kilometres away. The volcanic episode was short lived, and the heavy rainfalls and tropical climate has rapidly modified the surface expression of the volcano. The ejected ash has been strongly eroded and gullied with the

(a)   (b)

Figure 6.6 Post-volcanic features of Mount Pinatubo, Philippines, which last erupted in 1991. (a) Highly
degraded pyroclastic flows that have been covered by a dense tropical vegetation. (b) Crater lake
on the summit of Mount Pinatubo. The lake fills the caldera that formed by the collapse of the
central volcanic cone. The caldera walls expose a sequence of pyroclastic flows, volcanic breccias
and andesite lavas.

materials being carried as mud flows (lahars) for tens of kilometres beyond the base of
the volcano, tropical weathering has been intense, and new vegetation has rapidly grown
on the ash (Figure 6.6a). The summit caldera is now filled with tepid, non-toxic water
(Figure 6.6b).

The accumulation of volcanic material in general occurs close to volcanic vents, and their
distribution and thickness around such centres is largely dependent on the type of volcano and the pre-existing topography, in particular, where the flows follow the valleys (e.g.
Figure 6.5a and b) or in depressions related to faulting. Where the eruptions occur along
fissures, the erupted material, largely lava, may spread for great distances and may over
time establish plateaux formed by a thick sequence of successive lava flows (Figure 6.14a).
However, ash ejected high into the atmosphere (Figure 6.2) can be carried by winds for
many hundreds of kilometres from the volcanic centres deposited, in places, as blankets of
ash-fall tuff.

Because of the intermittent formation of primary volcanic material (lava, ash flows, air-
fall tuff), the changing location of the main and subsidiary vents, the intrusion of sills, the
development of weathered surfaces between volcanic episodes and the deposition of eroded
material on land and in water from the volcano, the vertical and lateral variations of volcanic
rocks encountered in cross-sections may be very complex. Figure 6.7 displays a schematic
cross-section through a typical volcanic sequence and illustrates some of the variations in
composition and geometries of rock units that may be anticipated in a volcanic environment. It emphasizes the caution that must be taken in the evaluation of any project area in

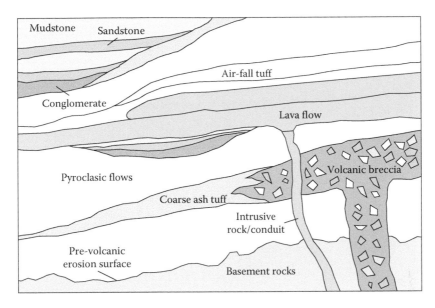

Figure 6.7 Schematic diagram of the interrelationships between different rock units in a hypothetical volcanic terrain.

volcanic terrains as the geotechnical properties of the components may change abruptly. Understanding the processes that are involved in the generation of a volcanic succession is the only way to establish a realistic geological model and thereby be able to predict the ground conditions to be expected.

Volcano shapes

The most recognizable shape of a volcano is an upright cone, for example Mount Fuji [35°22′N 138°42′E] in Japan or Shishaldin [54°45′N 163°56′W] in Alaska, United States. There, the volcanic materials build up on the volcano sides at their angle of rest and spread out only a relatively short distance from the volcanic vent. This material is in the form of either volcanic breccias, ash and lava flows or falls of ash, which are distributed fairly evenly around the vents as successive eruptions add to the volcanic edifice. Where the erupted material is largely lava and relatively non-violent, the volcano has a much lower profile and is referred to as a *shield volcano* after the shape of an upturned shield, for example Mauna Loa [19°28′N 155°36′W] in Hawaii, United States (Figure 6.3).

Calderas

Calderas form by the collapse of the central volcano into the void that developed by the extrusion of lava and tuff from the magma chamber below (Figure 6.8). Calderas are generally circular or elliptical in plan and bordered by sheer cliffs. They vary in width from a few hundred metres to many kilometres across and can readily be observed from satellite images, for example the active Kilauea caldera in Hawaii [19°28′N 155°36′W] or the extinct Ngorongoro Caldera (Figure 6.9) in Tanzania [3°10′S 35°34′E]. Eruptions within the caldera fill these basin-like features with thick sequences of layered volcanic material. For example, in Yellowstone Caldera, Western United States, lava covers about 350 km², and the dissected Jurassic caldera in Hong Kong (Sewell et al., 2000) accumulated ash-flow tuffs several hundreds of metres thick. Once the volcanic eruptions have ceased, calderas

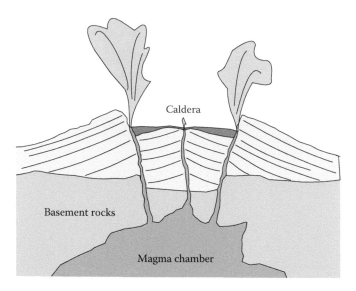

Figure 6.8 Schematic sketch of the development of a caldera. Subsequent cessation of volcanic activity and the erosion of the central volcano cones allow the formation of a crater lake.

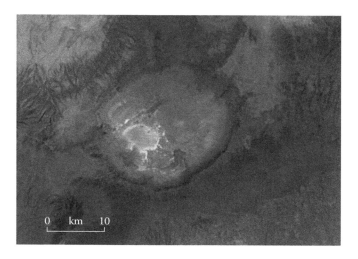

Figure 6.9 Google Earth, Landsat image of Ngorongoro Crater [3°10′S 35°34′E], an extinct caldera that was active some 2–3 million years ago, Tanzania.

can rapidly be filled with rainwater to form a lake. Mount Pinatubo in the Philippines has a caldera lake on its summit that has formed since the 1991 eruption (Figure 6.6b).

COMPOSITION OF VOLCANIC ROCKS

Volcanic rocks are primarily composed of an assemblage of crystals, volcanic glass, pumice, amygdales and exotic rock (lithic) fragments (McPhie et al., 1993). The form, composition and distribution of these components within the volcanic rock provide essential evidence of their history of formation, thereby allowing reliable assessments of the probable

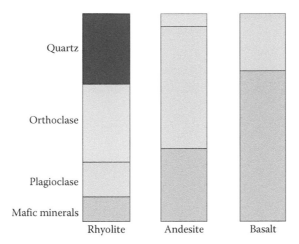

Figure 6.10 Simple classification of main extrusive igneous rocks based on the relative percentages of the main component minerals.

3D architecture of the various volcanic rock types to be made. The crystalline components of volcanic rocks are mainly silicate minerals, the most important of which are quartz, plagioclase (Ca, Na) feldspar, alkali (K) feldspar and ferromagnesian minerals (e.g. pyroxene, olivine), together with accessory minerals, including magnetite, iron sulphide and carbonate minerals (Figure 14.d). The classification and nomenclature of volcanic rocks has largely been based on the variation of percentage of these minerals (Figure 6.10), in particular the presence or absence of quartz and the relative amounts of the two feldspars. These variants in relative mineral percentages are related to the composition of the original magma, which in turn largely determines the type of volcanic eruption. The original magma would have also contained important amount of different gases, such as carbon dioxide and sulphur dioxide, together with water vapour, which would have partially controlled the melting point of the magma. It is the release of the gases as the magma reaches the Earth's surface that results in the explosive characteristics of some volcanoes.

Detailed classification of volcanic rocks has also been achieved by chemical analysis of both major and trace elements which have provided evidence for the origin and evolution of the magma. The variability of silica content, reported as the SiO_2 percentage within volcanic rocks, has been taken as one of the most widely used indicator elements. It varies from about 40% to over 80% which is reflected by the mineral content: quartz poor or absent (basalt) to quartz rich (rhyolite), respectively. Table 6.1 provides a simple trifold classification of the most common volcanic rock types with respect to their chemical composition and their plate tectonic setting.

Table 6.1 Basic threefold classification of volcanic rock types based on their SiO_2 values

Rock type	SiO_2%	Plate tectonic setting	Eruptions
Basalt	50	Mid-oceanic ridges, continental rift valleys	Lavas
Andesite	60	Continental margins associated with subductions	Lavas, airfall tuffs, pyroclastic flows
Rhyolite	70	Continental margins, magma contaminated with crustal material	Airfall tuffs, pyroclastic flows

The characteristic plate tectonic setting and the most common type of eruptive material are related to these values.

Crystal and crystal fragments

Large crystals grow at depth within magma chambers or volcanic conduits as the rock melt slowly cools. When the volcano erupts at the Earth's surface, these crystals are carried by the upwelling magma and either incorporated into lava flows and encased in a mosaic of much smaller crystals as the rock solidifies or ejected with ash in pyroclastic flows or into the air (Figure 6.11). Large crystals in lava flows, of which feldspar and quartz are the most common, are called *phenocrysts* and commonly have near-perfect crystal form. Early-formed crystals are also incorporated into violent volcanic eruptions that deposit volcanic breccias, pyroclastic flows and air-fall tuffs (Figure 6.11a). Crystals may retain their form, but more commonly they become fragmented due to the sudden pressure release, broken by collision with other crystals or embayed due to later resorption in the extreme heat of the lava or pyroclastic flow (Figure 6.11b). Minute crystal fragments also become part of the ash clouds that rise into the atmosphere.

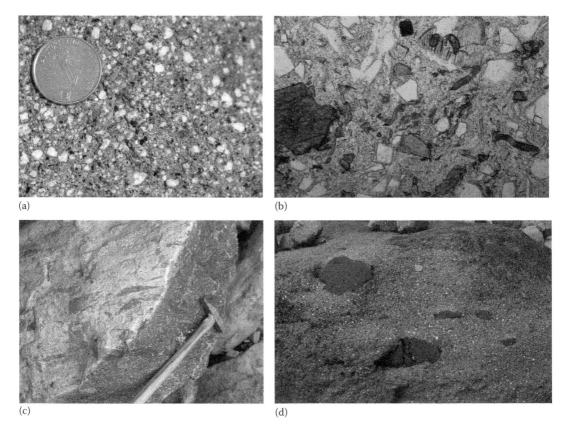

(a)   (b)

(c)   (d)

Figure 6.11 Typical components of explosive extrusive igneous rocks. (a) Crystals of white feldspar, grey quartz and black hornblende in a coarse-ash tuff, Hong Kong. (b) Thin section of a pyroclastic flow with ragged crystals of hornblende and quartz within a glassy matrix. Plane polarized light. View width 4 mm. (c) Lenses of flattened pumice (fiamme) within a fine-ash vitric tuff. The fiamme are observed best on the weathered surface, Ap Lei Chau, Hong Kong. (d) Sub-angular fragments of lava stripped from the walls of the volcanic conduit within a coarse-ash crystal tuff, Mount Pinatubo, Philippines.

Volcanic glass

Quenching of magma results in the formation of volcanic glass, with the rapid cooling being so quick that only skeletal or other poorly formed crystals can develop. Volcanic glass has a distinctive conchoidal fracture (similar to flint) and a glassy lustre. The most well-known variety of volcanic glass is the very silica-rich black or dark grey obsidian, which has been used since prehistoric times for implements and in jewellery. Small shards of volcanic glass commonly make up a large percentage of the fine-grained fraction of pyroclastic deposits and are formed by the explosive fragmentation of the magma (Figure 6.11b).

In highly explosive eruptions, a very vesicular-rich variety of volcanic glass, called *pumice* (Figure 6.14b), is formed that is commonly less dense than water and therefore can drift across oceans.

Fiamme (flame structures)

Fiamme are glassy, disc-shaped lenses within pyroclastic deposits that have flame-like shapes with wispy terminations (Figure 6.11c). They are formed by the compaction, flattening and welding of pumice fragments during the solidification of pyroclastic flows. Fiamme commonly impart a distinct foliation to the volcanic rock that parallels the bedding within the volcanic sequence. This foliation has been referred to as a *eutaxitic foliation* and has been used to interpret the orientation of thick pyroclastic flows where no contacts or other bedding features are exposed. For example, in Case Study 6.1, a fatal landslide occurred in weathered pyroclastic rocks. There, the eutaxitic foliation had been folded, and it was observed that where the eutaxitic foliation was vertical, the weathering was able to penetrate to the greatest depth. Instability of the thick weathering profile resulted in failure of the hill slope at this specific locality.

Rock fragments

Fragments of rock plucked from the walls of the volcanic conduits become incorporated into the material ejected from a volcano. The rock fragments, commonly referred to as *lithic fragments*, may be sourced from basement rocks, for example limestone or schist, below the volcano or from volcanic rocks erupted during an earlier episode of volcanism. Lavas generally do not contain lithic fragments due to the less explosive nature of the eruptions. However, in places, the chilled margins of lava flows and along the sides of volcanic conduits may break away and be enveloped by the magma. The lithic fragments in pyroclastic deposits are angular, except where partially resorbed, and the largest ones are more prevalent close to volcanic centres (Figure 6.11d). Lithic fragments can also be incorporated into the base of a pyroclastic flow as it moves over the ground.

Volcanic bombs may also be regarded as lithic fragments; however, they are rarely preserved in ancient volcanic rocks. The bombs display a characteristic 'bread-crust' texture to their surfaces; due to the rapid cooling, the ejected magma blobs (Figure 6.12a). Where they land, it is sometimes possible to observe that they have indented the layering in the pre-existing strata and are overlain by later deposits (Figure 6.12b).

LAVAS

Lava is extruded slowly at the Earth's surface either along fissures or focussed at one or more centres. Fissure eruptions account for vast outpourings of magma that over many cycles may cover thousands of square kilometres, for instance those of the Deccan Plateau basalts of India and the Giant's Causeway of Northern Ireland (Figure 6.13a). These initially fill the

(a)

(b)

Figure 6.12 Volcanic bombs. (a) 'Bread-crust' texture to the surface of a volcanic bomb, Mount Pinatubo, Philippines. (b) Volcanic bomb which fell into an ancient sequence of tuffs and sediments, St Andrews, Scotland.

topographic hollows but after many eruptions form planar surfaces and as a consequence are called *plateau basalts*. Individual lava flows may attain thickness of over 500 m, for example the Cerros de Chau lava flow in northern Chile, but ones of about 10–30 m are more common.

Lava surfaces

Lavas become more viscous and slower moving as they cool on exposure to the atmosphere. A skin develops on the upper surface of the lava flow that wrinkles or breaks up into blocks as the lava continues to move beneath it. Ropelike features caused by continued movement of the solidifying lava commonly cover the tops to many lava flows (Figure 6.13b), whereas in less viscous lavas, the solidified surface breaks up into blocks.

Pillow basalts

When hot lava is extruded from a vent under water, it solidifies quickly and a bubble-shaped crust is formed that expands into a pillow shape as more lava is forced upwards. Eventually, the pillows become solidified and the lava finds another way to escape – so forming the next pillow. The pillows have concave bases, cooling layers around their margins, radial cracks and in places central cavities (Figure 6.13c). Pillow lavas provide conclusive evidence for the environment of extrusion of the magma and an indication of the plate tectonic setting of ancient volcanic sequences.

Vesicles

The dissolved gases in the magma are expelled as the confining pressure is released on the eruption of the lava. The bubbles of gas are either ejected from the upper surface of the lava flow or entrapped as cavities, called *vesicles*, in the main body of the lava (Figure 6.14a). The shape of the vesicles is commonly spherical, but in some circumstances, they may become deformed during the on-going movement of the magma. Subsequent percolation of hot water through the lava flow results in the crystallization of minerals, such as quartz (agate) and calcite, within the vesicles. Partially or totally filled vesicles are referred to as *amygdales*.

Figure 6.13 Features of lava flows. (a) Sequence of multiple tertiary lava flows, separated by iron-stained weathered horizons. Lava flows display columnar jointing, Giant's Causeway, Northern Island (Courtesy of the Northern Island Geological Survey.) (b) Recent lava flow displaying a rubbly top that passes downwards into vesicular basalt. The flow has been overridden by a later flow that displays a ropy texture – Mount Etna, Sicily. (c) Precambrian pillow basalts that have been rotated into the vertical position during a tectonic episode. The pillows display chilled margins, radial cracks and concave lower surfaces – Llanddwyn Island, Wales. (d) Thin section of tertiary basalt composed dominantly of plagioclase feldspar laths (striped grey and white), olivine crystals (bright interference colours) and interstitial volcanic glass (black) – Scotland. Crossed polarized light, view width 4 mm. (Courtesy of the British Geological Survey.)

PYROCLASTIC DEPOSITS

Pyroclastic deposits are formed by the explosive eruption of material into the air from a volcano and are composed of broken fragments of rock, volcanic dust, pumice, crystals, glass shards and lava. Such deposits are also called tuff and this general term is in common usage. Many pyroclastic deposits are associated with the classical cone-shaped volcanoes on land (Figure 6.2), but may also form new islands offshore, for example the island of Surtsey off the coast of Iceland, which first erupted in 1964. The pyroclastic deposits range from large angular tuff breccias and lapilli tuffs within and close to the vent of the volcano (Figure 6.16a), ash-flow tuffs on the sides of the volcano and across the surrounding areas, to very-fine-grained air-fall tuffs, which may be found at a great distance from the volcanic centre.

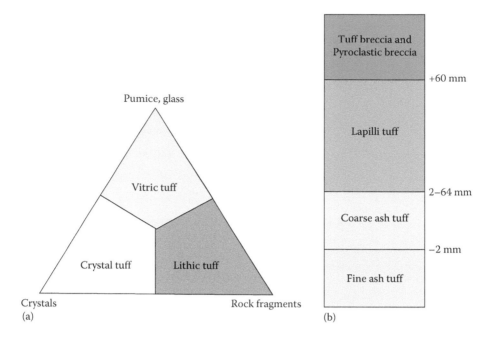

Figure 6.14 Classification of pyroclastic rock. (a) Composition of the main components. (After Le Maitre, R.W., eds., *A Classification of Igneous Rocks and Glossary of Terms. Recommendations of the International Union of Geological Sciences on the Systematics of Igneous Rocks*, Blackwell Scientific Publications, Oxford, U.K., 1989, 193pp.) (b) Grain size of fragments based on Fisher and Schminke (1984).

Pyroclastic deposits are classified according to the composition of the component fragments and their grain size. The dominant fragments in pyroclastic deposits are crystals, commonly broken, formed in the magma prior to eruption, glass and pumice composed of magma quenched on exposure to the atmosphere and rock fragments striped off the walls of the volcanic conduit as the magma was forced to the surface through pre-existing rocks. The percentages of these different types of fragment are used in naming the pyroclastic deposits as crystal, vitric or lithic tuff (Figure 6.14a). The size of the volcanic fragments also has been used to further classify tuffs into tuff breccia, lapilli tuff and ash tuffs (Figure 6.14b), and the relative percentage of these size fractions is an additional component in determining the name of the pyroclastic deposit.

In site investigation boreholes through pyroclastic rocks, it is rare to identify contacts between the different deposits, because these contacts are commonly gradational or have been obscured by weathering. It is therefore essential to obtain as much information as possible using the tuff classification earlier and any structures that may be present, such as a eutaxitic foliation, in order to construct the geometry of the deposits. Figure 6.15 displays a range of pyroclastic tuffs taken from site investigation boreholes in Singapore and Hong Kong.

Volcanic breccias

Most volcanic breccias are formed during two processes: explosive eruptions that strip fragments of country rock from the volcanic conduit or fragmentation of the volcanic

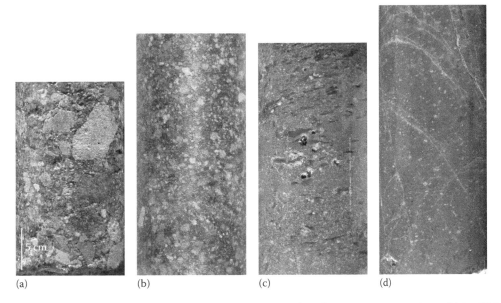

Figure 6.15 Examples of pyroclastic rocks from core samples taken during site investigations. (a) Lapilli lithic tuff with rock fragments of volcanic and sedimentary rocks. Note angular form to the fragments and chlorite-rich matrix, Singapore. (b) Coarse-ash crystal-lithic tuff, Hong Kong. (c) Fine-ash lithic tuff. Pumice fragments have been altered to green chlorite, Singapore. (d) Fine-ash vitric tuff, with scattering of small feldspar crystals, Hong Kong.

rock by quenching, mass wasting or other near surface processes during or just after the volcanic eruption. In the former case, tuff breccias were deposited within or close to volcanic vents and generally have limited lateral distribution. Commonly, they have circular to elliptical shapes and sharp sub-vertical contacts. They contain large, commonly unsorted, angular to sub-rounded, lithic blocks (rock fragments) greater than 60 mm derived from the walls of volcanic vents. As a result, the composition of the blocks will depend on the depth from which the blocks were sourced: they may be all of one rock type, for example volcanic blocks from an earlier volcanic eruption (Figure 6.16) or a wide variety of rock types sampled from the country rocks (e.g. sandstone, limestone) underlying the volcano.

Rapid quenching of a lava flow on entering water or magma at the cool margins of a subvolcanic intrusion results in the *in situ* fragmentation of the rock and formation of volcanic breccias. The brecciated blocks are generally angular and show little signs of movement in that they have a jigsaw appearance. Such breccias commonly form narrow selvedge adjacent to the primary volcanic rock; however, at the surface they may be eroded and redeposited in a sedimentary environment.

Ash-flow tuffs

Ash-flow tuffs avalanche down the sides of the volcano and can extend many kilometres away from the volcanic centre where the flows are constrained by the pre-existing topography. They keep close to the ground surface and attain speeds of up to 200 km/h and in part can reach over 1000°C. Successive ash flows can accumulate in topographic lows formed by erosion or collapse of the central part of the volcano, for example the pyroclastic tuffs of the eastern part of Hong Kong (Sewell et al., 2000). In these cases, the ash-flow deposits may

Figure 6.16 Volcanic breccia formed of unsorted blocks of coarse-ash tuff set in a slightly lighter-coloured and finer-grained ash groundmass, Ap Lei Chau, Hong Kong.

attain hundreds of metres in thickness. Ash flows can also maintain their integrity when they enter water or are ejected subaqueously. When the avalanche comes to rest on the lower slopes of the volcano, the fragments commonly become welded together, the glass becomes devitrified and the pumice flattened, resulting in the characteristic textures of welded tuffs, including glass shards and fiamme. In the unwelded state, ash-flow tuffs are weak, whereas welding makes such pyroclastic deposits extremely strong.

Air-fall tuffs

The lighter material is ejected high into the atmosphere, at times well over 10,000 m into the air, forming the characteristic plume, for example above Mount Mayon in the Philippines (Figure 6.2). The ash mostly falls close to the volcanic centre and may blanket the surrounding areas, but depending on the strength of the wind, it may be carried for hundreds of kilometres to settle as thin ash layers. For example, air-fall ash from Mount Pinatubo volcano was carried to Singapore during the 1991 eruption, and more violent eruptions, such as Krakatau, Indonesia, in 1883 sent dust-sized ash particles into the upper atmosphere that gave spectacular sunsets around the Earth for 2 years and lowered the global temperature by 1.2°C.

STRUCTURES WITHIN HOT VOLCANIC ROCKS

The internal structures of lavas and pyroclastic deposits display a spectrum of internal structures related to their mode of emplacement, the viscosity of the magma and the cooling conditions. On the other hand, the most internal structures observed in units of air-fall tuff and volcanic breccia are not formed directly by the crystallization of magma but are controlled by the density and size of the fragments. The only exception to this is when pyroclastic deposits accumulate in great thicknesses, the cooling history of the whole unit is not dissimilar to lava flows and columnar joints may develop (Figure 6.19b).

Figure 6.17 Bedding features in volcanic rocks. Fragmented blocks of lava and volcanic ash form an ill-defined layering (bedding) on the flanks, a cinder cone, Mount Etna, Sicily.

Bedding features

Within a volcanic cloud, there is a natural separation of the different particle sizes and weight fractions, so that in the case of air-fall tuffs, the finest-grained fraction will slowly fall to the ground and will be carried by wind further from the volcanic centre. Thus, many air-fall tuffs are finely bedded with the layers from each eruption displaying a grading from the heavier material at the base of the bed to the lightest at the top. In addition, each eruptive phase will produce slightly different material so that bedding in an air-fall tuff unit can be defined by colour and/or particle size.

Close to the volcanic vents, the larger ejected fragments form breccias that, for the most part, are chaotically arranged and unsorted. However, in places, there is also a slight layering developed due to the different weights of the fragments that fall back into the vent or accumulate on the slopes adjacent to the vent. Some slopes preserve a layering due to different intensities and compositions of the successive eruptions. For example, one of the satellite cones on Mount Etna, Sicily, displays an ill-defined bedding that parallels the slope of the cone of the ejected material that includes fragments of lava, coarse-ash and volcanic bombs (Figure 6.17).

Cooling joints

Cooling of hot volcanic deposits commonly results in the formation of contraction joints, which have the form of elongate prisms with dominantly hexagonal cross-section. They are most common in lava flows (Figures 6.13a and 6.18a) but can also occur in thick welded pyroclastic flows where cooling of the ash has been slow (Figure 6.18b). There appears to be little compositional variation from the sides of the columns to their centres. The columns are orientated perpendicular to the contact with the pre-existing rock along the base of the flow, or the top exposed cooling surface. Some lava flows show a vertical zonation of the columns from a well-formed colonnade at the bottom of the flow with the columns oriented perpendicular to the base of the flow, overlain by an entablature of unorientated shorter and thinner columns

(a)

(b)

Figure 6.18 Cooling joints. (a) Columnar cooling joints in the basal half of a thick basalt lava flow, Fingal's Cave, Isle of Staffa, Scotland. The majority of the columns are hexagonal in cross-section and orientated at right angles to the margins of the flow. The entablature composed of chaotic short basalt columns overlies the colonnade. (b) Cooling joints in a sequence of pyroclastic flows. Note that the majority of the columns have hexagonal cross-sections, Ninepin Islands, Hong Kong.

and topped by smaller columns perpendicular to the upper cooling surface (Figure 6.18a). In a few places, the columns may display radial patterns, may be curved or display complex patterns. These forms can be due to local movements in the underlying lava, deformation of the columns prior to final solidification of the lava or nonplanar cooling surfaces.

Volcanic foliation

Foliation structures within volcanic rocks are most commonly formed in welded pyroclastic flows or in the more viscous (SiO_2-rich rhyolite) lavas. They are formed either by laminar flow within the eruptive unit or by compaction and welding of the pyroclastic material mostly after the flow has come to rest. In pyroclastic flows, the formation of flattened fiamme imparts a fabric to the rock called a *eutaxitic foliation* (Figure 6.19a). In rhyolitic lavas, the foliation is defined mainly by variations in grain size, slight compositional differences and colour that developed during the laminar flow of the viscous magma. Deformation of this foliation during movement of the lava commonly produces disharmonic fold patterns that in cases where the folds are asymmetric can indicate the original direction of flow of the rhyolitic lava (Figure 6.19b).

LAHARS

Lahars are mudflows composed mostly of ash that sweeps at great speed down the sides of volcanoes shortly after their eruption. When the lahars follow valleys, they can travel for many

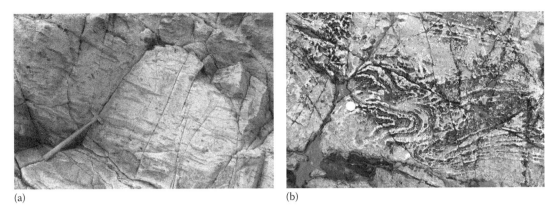

(a)   (b)

Figure 6.19 Foliations within hot volcanic rocks, Hong Kong. (a) Eutaxitic foliation in an ash-flow tuff, defined by the parallel alignment of the fiamme. (b) Folded flow foliation in rhyolitic lava. The asymmetry of the folds indicates that lava flowed towards the right of the picture.

tens of kilometres from their source – as a result, they are extremely dangerous and have caused catastrophic devastation and loss of life many kilometres from the volcanic centre, burying the whole town in a thick blanket of mud, trees and rock debris. The water in the mudflows generally comes from heavy rainstorms or breached caldera lakes. Although lahars are most frequent immediately after the deposition of the pyroclastic deposits, when the ash is not consolidated nor the ground surface stabilized by vegetation cover, they can occur many years later than the volcanic eruption when erosion has exposed the tuffs in gullies and canyons. For example, the lahars associated with the 1991 Mount Pinatubo eruption in the Philippines covered the town of Los Angeles, some 20 km west of the volcano to a depth of 10 m, with much loss of life. Some 20 years later, lahars still flow down the valleys (Figure 6.20a), although with not so much frequency or force, deriving their material from erosive channels in sequences of loose tuff and previous lahars that flank the volcano (Figure 6.20b).

Recognition of lahar deposits that formed in the geological past provides evidence for the style of volcanic eruption that took place, the climate at that time and an indication of the geometry of the rock units. For example, a 4 m thick tuffaceous mudstone containing randomly distributed angular rock boulders exposed in a recent rock cut in Turkey has been interpreted as an ancient lahar (Figure 6.20c).

FUMAROLES

During the later stages of a volcanic cycle, it is common for gases to be emitted from vents, long after any lava or pyroclastic flows have erupted. These gases when mixed with water are periodically ejected as explosive *geysers* (Figure 6.21a). Groundwaters are also heated by the circulation of the water deep within the Earth's crust close to the underlying magma chambers. Such geothermal hot waters (hydrothermal water) are enriched with the elements silica, calcium, iron and sulphur, together with a multitude of minor elements such as arsenic, copper and even gold. When these mineral-enriched waters flow slowly, but continuously, onto the ground surface, the elements are precipitated as encrustations of mainly travertine (calcium carbonate) (Figure 6.21b) and siliceous sinter (cryptocrystalline quartz) (Figure 6.21c).

(a)

(b)

(c)

Figure 6.20 Lahars. (a) Recent lahar following a valley to the east of Mount Pinatubo, Philippines. The main drainage pathways for the lahars are well seen on the satellite image (Figure 6.5c). (b) A sequence of lahars formed during the eruption of Mount Pinatubo in 1991 and exposed in a cliff formed by recent river erosion. (c) Probable Quaternary lahar deposits interbedded with lake deposits. Note that the angular blocks of volcanic rock are randomly supported in a fine-grained, tuffaceous mudstone groundmass, Aliaga, Turkey.

(a)

(b)   (c)

Figure 6.21 Examples of hot springs and geysers from Tibet. (a) Geyser in an active hot spring area. (b) Accumulation of travertine sinter in cascading ponds fed by low volumes of geothermal waters. (c) Layered sinter from old hot springs. The deposits are rich in iron sulphides, which have been oxidized across the low hill in the distance.

CASE STUDY 6.1   WEATHERED TUFF: FATAL LANDSLIDE, SHUM WAN, HONG KONG

[22°14.4′N 114°10.0′E]

In August 1995, one of the largest landslides to affect Hong Kong over the previous 25 years occurred at Shum Wan, near Aberdeen on the south side of Hong Kong Island (Figure CS6.1.1). During an intense rainstorm, approximately 26,000 m³ of soil detached as a semi-coherent slab down the hillside into the harbour. The hillslope was heavily wooded at the time of the failure, and the rock outcropped along Sham Wan Road that runs along the old foreshore. There were two fatalities and the shipyards were destroyed as a result of the landslide.

GEOLOGY

The hillside is underlain by folded lapilli-bearing fine-ash vitric tuffs (Figure CS6.1.2). These display a prominent eutaxitic foliation that parallels the orientation of the primary bedding in volcanic rocks (Kirk et al., 1997). The folds in the region are south-east trending and asymmetric, with steeply dipping north-eastern limbs (Figure CS6.1.3). Within the zone of failure, the foliation is sub-vertical, whereas on its flanks, the foliation is low dipping (Figure CS6.1.4). At least two sets of steeply dipping joints cut the volcanic sequence, and close to the ground surface, sheeting joints dip shallowly towards the west out of the hillside. On the wooded hillside, above Shum Wan Road, a variable thickness of *in situ* soils were present – the result of tropic weathering over an extended period of time. Erosion had largely removed these soils from the hillside, but at the landslide site, they were over 15 m thick. The soils consisted of Grade IV and Grade V soils and sharply overlay Grade III rock. Close to the base of the soil profile, pure white kaolinitic clay had been deposited within relict joints, and just above the bedrock interface, a persistent layer of buff kaolinitic clay,

Figure CS6.1.1 Shum Wan Road landslide prior to extensive remedial works and slope revegetation.

Figure CS6.1.2 Shallowly dipping eutaxitic foliation in vitric tuff, south coast of Ap Lei Chau.

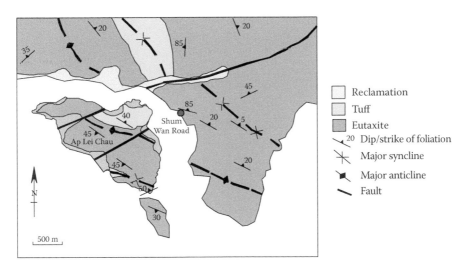

Figure CS6.1.3 Main structural geological features of the Ap Lei Chau area, southern Hong Kong Island, showing orientation of eutaxitic foliation and traces of fold axial surfaces. (Based on Geotechnical Engineering Office, 1987)

up to 100 mm thick, had developed (Figure CS6.1.5 and CS6.1.6). The buff clay contains weathered fragments of volcanic rock, and during downslope movements, the clay was forced downwards into wedge-shaped cracks in the underlying substrate. The low friction angles (18° or less) of the clay and elevated water pressures above the clay layer during an intense rainstorm resulted in the detachment of the soil and weathered rock slab.

The location of the landslide was controlled by the geometry of the folds in the volcanic sequence and the consequential increased depth of tropical weathering along the steeply

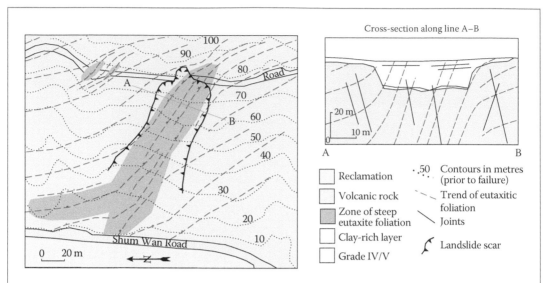

Figure CS6.1.4 Geology of the Shum Wan Road landslide. (Based on Geotechnical Engineering Office, 1996.)

Figure CS6.1.5 Cut block of the landslide rupture surface showing buff and white, clay-filled wedges and joints in weathered tuff, overlain by landslide debris. Width of block 25 cm. (From Fyfe et al. 2000.)

dipping fold limbs. Water preferentially flowed into the rock along the sub-vertical foliation planes, thereby enhancing the decomposition of the volcanic rocks and lowering the overall depth of weathering. Within this zone, the soil/rock interface was largely controlled by low-dipping sheeting joints, whereas the steeply dipping tectonic joints acted as back-release surfaces. All these factors led to a localized thick soil carapace becoming unstable during extreme weather conditions.

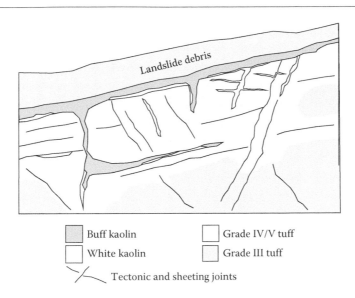

▨ Buff kaolin	▢ Grade IV/V tuff
▢ White kaolin	▢ Grade III tuff

Tectonic and sheeting joints

Figure CS6.1.6 Schematic representation of the main features of the landslide at Shum Wan. (After Fyfe et al., 2000.)

ENGINEERING CONSIDERATIONS

- Sub-vertical foliation in the volcanic tuffs largely controlled the location of deep weathering zones across the hillside.
- A clay layer developed at the base of soil profile as a result of slope movements.
- The clay layer has very low friction angles and adversely dips out of the hillside.
- Tectonic joints acted as back-release surfaces.

CASE STUDY 6.2 IGNEOUS AND SEDIMENTARY ROCKS: BRIDGE FOUNDATIONS, RIVER FORTH, SCOTLAND

[56°0.3′N 3°25.1′W]

T. Berry, J. Brown, T. Casey and P. Mellon
Employers Delivery Team

Owing to the condition of the existing Forth Road Bridge, the Scottish government concluded that a replacement crossing was required to safeguard this vital connection across the Firth of Forth. The Queensferry Crossing is a cable-stayed bridge, comprising a 2.67 km crossing supported by three towers and approach viaducts, west of the existing bridges (Figure CS6.2.1). As part of the preparatory works, an extensive ground investigation was carried out over 4 years to investigate the ground conditions along the proposed alignment. The ground investigations revealed important information about the geological history of the area, greater understanding of the geographical extent of anticipated geological formations and variation of composition and geotechnical properties of strata encountered. Ground investigations and development of the geological model for the Queensferry Crossing have redefined the geology beneath the Firth of Forth.

Figure CS6.2.1 The new Queensferry Crossing under construction with the 1964 Forth Road Bridge and the 1890 Forth Rail Bridge in the background.

GEOLOGY

The site (Figure CS6.2.2) is within a geological area known as the Midland Valley, defined to the north by the Highland Boundary Fault and to the south by the Southern Upland Fault and developed late in the Silurian Period (416–443 Ma). The bedrock comprises Carboniferous sedimentary strata, formed in a series of marine transgressions and regressions, with various igneous intrusions, emplaced following the Variscan orogeny (380–280 Ma). The site is also cut by major east-west-trending fractures and extensional faulting. An eastward-flowing drainage pattern developed in the Neogene Period (23–2.6 Ma), and the form of the Forth Valley was subsequently modified during the Quaternary Period (2.6 Ma to present) by glaciations, with significant glacial deposits burying the bedrock. Relative changes in sea level due to the glaciations and subsequent isostatic rebound led to the formation of raised beaches that are characteristic of the area.

The geology encountered during the ground investigations, which incorporated extensive and specialized *in situ* and laboratory testing, was generally in accordance with that anticipated from the preliminary studies. However, through intrusive investigations, the understanding of the geology was further developed by the discovery of the following key pieces of geological information in the five main areas of the site:

- Identification of highly altered alkali dolerite ('white trap'), with variability in alteration intensity (Figure CS6.2.3)
- Reclassification of sedimentary strata in the central marine area
- Identification of relative timing of igneous intrusions into partially and fully lithified sedimentary strata (Figure CS6.2.4)
- Identification of volcanic neck features associated with volcaniclastic deposits
- Mapping of previously unidentified superficial deposits

Significant variations of intact strength of the strata encountered were observed, particularly in horizons and zones which exhibited evidence of alteration associated with emplacement of dolerite intrusions. In places, the alkali dolerite has been significantly altered into an aggregate of

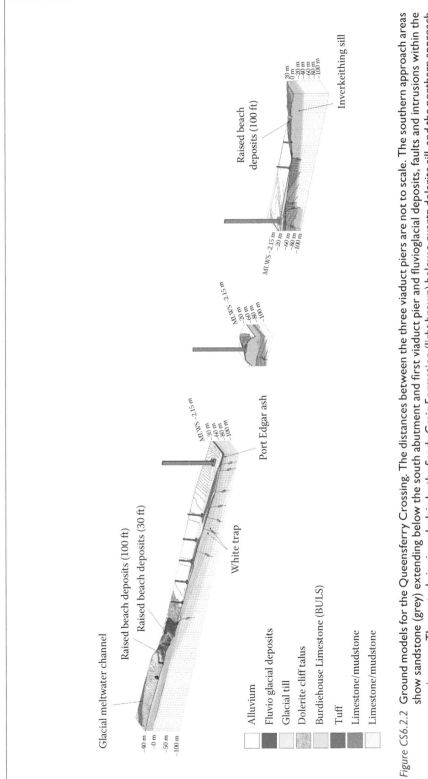

Figure CS6.2.2 Ground models for the Queensferry Crossing. The distances between the three viaduct piers are not to scale. The southern approach areas show sandstone (grey) extending below the south abutment and first viaduct pier and fluvioglacial deposits, faults and intrusions within the marine area. The central pier is underlain by the Sandy Craig Formation (light brown) below a quartz dolerite sill, and the northern approach areas have similar geology with the addition of a volcanic tuff vent and upper and lower fluvioglacial deposits.

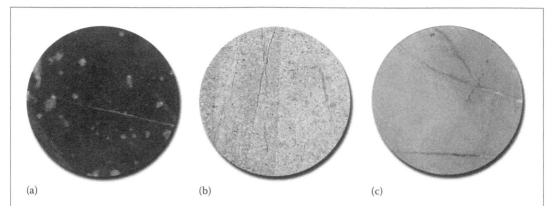

(a)   (b)   (c)

Figure CS6.2.3 Polished core showing progressive alteration of dolerite intrusions, Diameter of core 100 mm. (a) Unaltered dolerite. (b) Partially altered dolerite. (c) Intensively altered dolerite (white trap).

Figure CS6.2.4 Complex igneous contact between the mudstones of the Sandy Craig Formation (left) and the altered dolerite (right). Diameter of core 100 mm.

calcium, magnesium and iron carbonates with kaolin and muscovite, referred to in the published literature as 'white trap' (Figure CS6.2.3).

The volcaniclastic deposits (tuff) encountered in the northern marine area were observed to exhibit variations of compressive strength in relation to changes of moisture content following time since sampling. Special measures to preserve samples for testing rapidly following extraction and logging were implemented.

ENGINEERING CONSIDERATIONS

- The confirmation of rockhead levels at the flanking towers was essential in determining the required span lengths for the main cable-stayed bridge. If the depth to rockhead had increased more significantly, as is the case in other areas of the Firth of Forth and many other glaciated valleys, a larger span may have been required.
- While coal, limestone and oil shale units were known to have been worked in the near vicinity of the crossing, no underground workings were encountered beneath the alignment of the main crossing, although workings were encountered below the sections of the network connections.
- The ground conditions determined from the site investigations allowed the methods of excavation and construction (i.e. dredging, caissons and cofferdams) to be developed.
- The material strengths determined allowed the foundations, which were mainly pad foundations constructed onto rock, to be sized.

Chapter 7

Metamorphic and hydrothermal rocks

Metamorphism is the process by which pre-existing rocks are altered by chemical reactions under changing temperature, pressure and shearing stress. The original mineralogy and fabric of both sedimentary and igneous rocks can be changed due to the effects of plate tectonic processes with the resultant increase in temperatures and pressures (dynamic metamorphism), the emplacement of younger hot igneous bodies (contact metamorphism) or the dispersion of relatively low-temperature fluids and gases, commonly mineralized, during the later stages of igneous activity (hydrothermal alteration).

Dynamic metamorphism can extend over many hundreds of square kilometres and as a consequence is sometimes referred to as regional metamorphism. Dynamic metamorphism is due to a combination of increasing pressures, temperatures and shearing stresses that range from very low following the deposition of the rocks (Figure 7.1a) to extremely high (Figure 7.1b) when the rocks can partially melt. Contact metamorphism on the other hand is very localized and restricted to narrow zones adjacent to igneous intrusions where pressures are low but temperatures reaching 900°C. Hydrothermal alteration is again limited to small areas but its effects can extend for several kilometres from and within the igneous source. Generally, the fluids have temperatures no more than 200°C. Where there are conduits for the hydrothermal fluids, this type of alteration can extend great distances from the original source of the fluids and gases, for example, along a fault zone.

In all cases, the mineralogy and the structure of the rock change in order for the rock to become in equilibrium with the superimposed physical conditions and influx of hydrothermal fluids. For a full description of metamorphic rocks, the reader is referred to standard textbooks (e.g. Fry, 1984; Geological Society, 1991; Fettes and Desmons, 2011).

ENGINEERING CONSIDERATIONS

Metamorphism can have a profound influence on the geotechnical properties of the original rock in respect to their strength, hardness, friability, fracture potential, weathering susceptibility and hydrogeological characteristics. An understanding of the regional geological setting and processes to be expected in any project area subject to either metamorphism or hydrothermal alteration is therefore essential in formulating a comprehensive assessment of the ground conditions to be anticipated. These will change depending upon the type and intensity of the metamorphism and hydrothermal alteration – they therefore must be assessed independently.

Dynamic metamorphism is associated with the development of foliations which impart planes of weakness in the rock. These are described and discussed in detail in Chapter 9. The geotechnical properties may change across a dynamically metamorphosed area as the grade increases. For example, slaty cleavage in low-grade metamorphic rocks may be very closely spaced and continuous providing significant planes of failure, schistosity in medium-grade

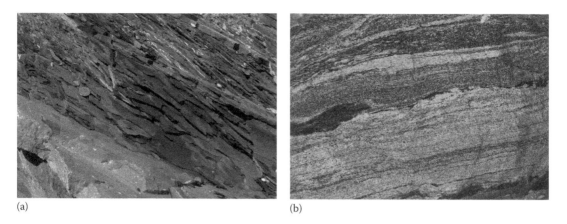

(a)   (b)

Figure 7.1 Dynamic metamorphism. (a) Low-grade metamorphosed mudstone with the development of incipient foliation, Aberystwyth, Wales. (b) High-grade, banded gneiss with dark, amphibole-rich lenses. The original rock was probably a layered sedimentary rock composed of sandstone, siltstone and impure limestone. View width 20 cm, Kerala, India.

rocks commonly is less planar but has low friction angles, and gneissosity in high-grade rocks can have planes of weakness along compositional bands or the foliation, but in places, the recrystallization of the gneiss has allowed it to act as a homogeneous, high-strength rock mass.

The geotechnical properties of different types of metamorphosed rocks vary considerably and must be assessed separately. The main geological features that could be important in any ground engineering project where dynamic metamorphism, contact metamorphism or hydrothermal alteration has taken place are given in the succeeding text under separate headings.

Dynamic metamorphism

- Foliation planes impart planes of weakness through the rock and these may have low friction angles due to the crystallization of micas.
- Low- and medium-grained, dynamically metamorphosed rocks are generally very susceptible to weathering.
- Strengths are dependent on the directions.
- Quartz veining, which can affect borings or driven piles, is commonly present.

Contact metamorphism

- In many cases, contact metamorphism increases the strength of weaker sedimentary or volcanic rocks.
- Adjacent to granites, calcareous rocks can be associated with cavitous marble and segregations of iron minerals.
- Commonly, there is complex interdigitation of rock types and mineral associations.

Hydrothermal alteration

- Chloritization may lessen the overall strength of the rock and reduce the friction angle of joints.
- Silicification greatly increases the hardness of the country or fault rocks.
- Greisenization may result in pockets of very weak, mica-rich ground.
- Kaolinitization of granitic rocks may result in wide tracts of extremely weak and easily weathered rock.

Case Study 7.1 describes the planning and construction of a road in east Nepal that traversed a wide range of metamorphic rocks from low-grade phyllites, through medium-grade schists, to high-grade gneisses. The variability of the metamorphic fabrics in these rocks was one the factors that had to be considered when assessing the stability of the steep hillslopes. The metamorphic fabric and mineralogy also largely controlled the quality and suitability of the rocks for construction materials. Case Study 7.2 details some of the geological features that needed to be addressed in the construction of a shaft and tunnel through high-grade metamorphic rocks in the United States. The reader is also referred to Case Studies 8.1, 9.1 and 10.1, which describe engineering projects that encountered metamorphic or hydrothermally altered rocks. Case Study 6.2 describes the hydrothermal alteration of a basic intrusion into a rock rich in carbonate minerals, clay and mica, which has significantly changed its compressive strength.

CONTACT METAMORPHISM

Contact metamorphism is caused by increased heat adjacent to intrusive igneous bodies (Figure 7.2). The width of the contact metamorphic zone is largely dependent on the size of the igneous body. Smaller igneous bodies such as dykes and sills can be very hot but cool rapidly and, as a result, are generally associated with narrow contact metamorphic zones, maybe less than a metre wide (Figure 7.3a), whereas larger igneous bodies, such as a granite pluton, cool more slowly and have affected larger areas. The mineralogy of contact metamorphic rocks is highly variable, and in places, large unorientated crystals may grow. For example, the high-temperature mineral andalusite (Al silicate) may crystallize in fine-grained sedimentary rocks that contain sufficient aluminium (Figure 7.3b).

Contact metamorphic mineralogy is controlled by the composition the host rock, which provides the elements that are required for the formation of new minerals. In some rocks, such as pure limestone and quartz sandstone, no new minerals formed during contact metamorphism; rather the calcite and quartz are recrystallized into a mosaic of interlocking crystals, which can greatly increase the strength of the original rock. The metamorphic mineralogy may change between adjacent sedimentary beds of slightly different composition (Figure 7.4a). For example, metamorphic biotite and amphibole are concentrated in the brown and greenish layers, respectively, that reflect the compositions of the original beds in an andesite tuff (Figure 7.4a). This compositional control is also well illustrated by reference a contact metamorphosed volcanic breccia from Hong Kong (Figure 7.5). Here, calcium-bearing

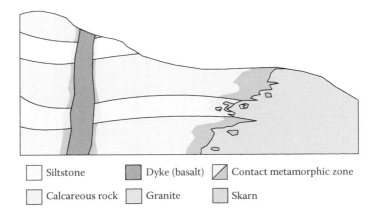

☐ Siltstone	▨ Dyke (basalt)	▨ Contact metamorphic zone
☐ Calcareous rock	☐ Granite	▨ Skarn

Figure 7.2 Schematic diagram of the development of contact metamorphic rocks.

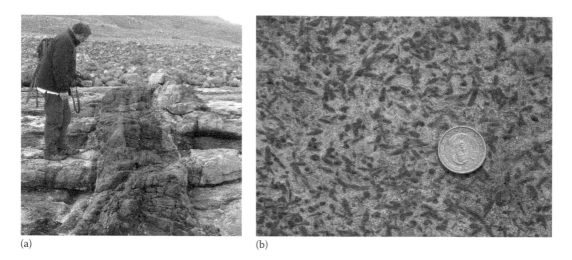

(a)   (b)

Figure 7.3 Contact metamorphism. (a) Narrow pink and greenish contact metamorphic zone in bedded limestone and shale adjacent to basalt dyke, Hebrides, Scotland. (b) Andalusite (Al silicate) crystals within metamorphosed mudstone adjacent to granite pluton, Portugal.

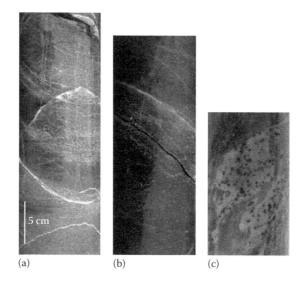

(a)   (b)   (c)

Figure 7.4 Contact metamorphism of bedded sedimentary and volcanic rocks from site investigation boreholes, Hong Kong. (a) Biotite and amphibole in a bedded volcanic tuff, preferentially concentrated within brownish and green layers, respectively. (b) Hard, dark green hornfels. (c) Dark, mineral spots within bleached areas in metamorphosed sandstone.

garnet has crystallized only within the impure limestone clasts, whereas the epidote (Ca, Al, Fe, silicate) is restricted to the volcanic matrix.

Fine-grained rocks such as mudstone are commonly metamorphosed to a hard, commonly dark-coloured, homogeneous rock called *hornfels* (Figure 7.4b), which commonly displays small mineral spots composed of either small clusters of hornblende and chlorite (dark green minerals) or feldspar (white mineral). The hornblende crystal clusters (spots) in a sample of core from very fine-grained sandstone adjacent to a granite pluton in Hong Kong (Figure 7.4c)

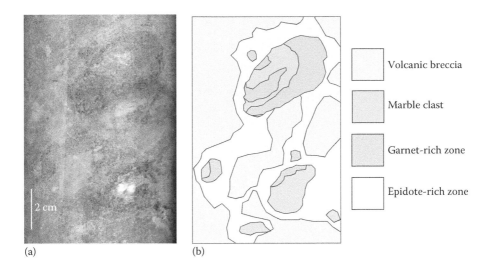

Figure 7.5 Contact metamorphism of a volcanic breccia, Tuen Mun, Hong Kong. (a) Core sample of a volcanic breccia. Garnet has crystallized within the limestone clasts, whereas the crystallization of epidote is restricted to the volcanic matrix. (b) Graphic representation of the adjacent core sample.

are associated with the bleaching of the host rock due to the removal of iron and magnesium minerals from the adjacent rock. Impure limestones are metamorphosed to calcsilicate rocks in which iron/calcium silicate minerals, such as garnet, epidote and pyroxene, are formed. For example, a contact metamorphism zone some 20 cm wide has developed along the margins of a basalt dyke cutting bedded limestone in the Hebrides of Scotland (Figure 7.3a). The zone contains calcsilicate minerals that give the rock its pinkish colour.

DYNAMIC METAMORPHISM

Dynamic metamorphism is associated with increased hydrostatic pressure, temperatures and shearing stress caused by plate tectonic processes over large areas or in more restricted zones close to faults. Collision or subduction of continental plates provide the mechanism for thick piles of crust (old sedimentary, igneous and metamorphic rocks) to be stacked and folded into mountain chains or to be taken down into the deeper parts of the earth's crust. The depths at which regional metamorphic takes place vary from about 5 to over 30 km. As the rocks are subjected to increasing pressures and temperatures, the original minerals in the rock become out of chemical equilibrium, and a sequence of new minerals are formed which define the grade of metamorphism.

At the lowest metamorphic grades, the clays in the mudstone and fine-grained sandstone are recrystallized and start to define an incipient foliation (Figure 7.1a). This becomes more defined as the grade increases in response to differential confining pressures and elevated temperatures. At this stage, the rock takes on a *slaty cleavage* that defines the dominant discontinuity of the rock (Figure 9.14b and c). With increasing temperature and pressure, the slate passes into a phyllite in which the white micas are larger and coat the foliation planes to give the rock a shiny appearance (Figure 7.7a). In medium-grade rocks, it is common for the rocks to develop a wavy foliation (schistosity – see Chapter 9) defined by platy mica flakes and elongation of the quartz and feldspar crystals (Figure 7.6a). New minerals

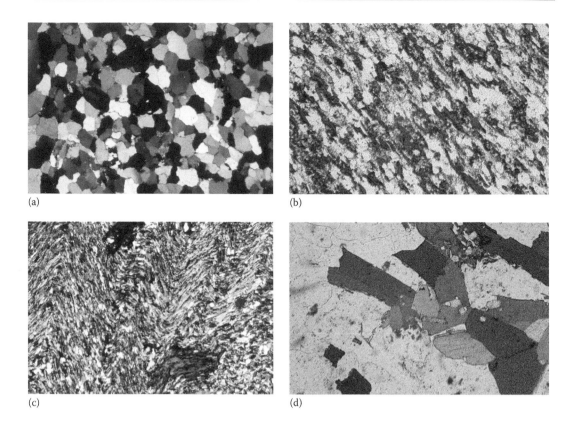

Figure 7.6 Thin sections of dynamically metamorphosed, quartz- and clay-bearing, sedimentary rocks from Scotland. (a) Quartzite with granular, intergrown quartz crystals. Crossed polarized light, view width 3 mm. (Courtesy of the British Geological Survey.) (b) Schist with brown biotite flakes defining a penetrative, slightly wavy foliation. Plane polarized light, view width 3 mm. (c) Crenulated schist with muscovite flakes concentrated in late-stage foliation. Crossed polarized light, view width 5 mm. (Courtesy of the British Geological Survey.) (d) Gneiss with elongate, greenish brown biotite flakes and prismatic, green hornblende crystals concentrated within the darker layer of a banded gneiss. Plane polarized light, view width 5 mm.

within schists characterize a series of metamorphic zones, each of which reflects a particular range of temperatures and pressures. The metamorphic zones, in order of increasing temperature and pressure, are defined by the first appearance of chlorite, biotite, garnet, staurolite, kyanite and sillimanite. The schist of the kyanite metamorphic zone displayed in Figure 7.5b contains large crystals of pink garnet and blue kyanite set in a quartz and feldspar matrix. At the highest grades of metamorphism, the rocks approach their melting point and there is commonly a differentiation of the minerals into bands. These rocks are referred to as gneisses and display a wide variety of textures and fabrics dependent on the tectonic environment and the original composition of the rock. In banded gneisses, there is a differentiation into layers rich in dark-coloured minerals (mafic minerals), such as hornblende and pyroxene, and those rich in lighter-coloured minerals such as quartz and feldspar (Figure 7.7c). In thin section, the coarse hornblende crystals are seen to be partially aligned parallel to the banding (Figure 7.6b). These amphibole-rich banded gneisses could have originated from volcanic rocks. Some granitic rocks develop large lenses or augen of feldspar at the highest grades of metamorphism (Figure 7.7c).

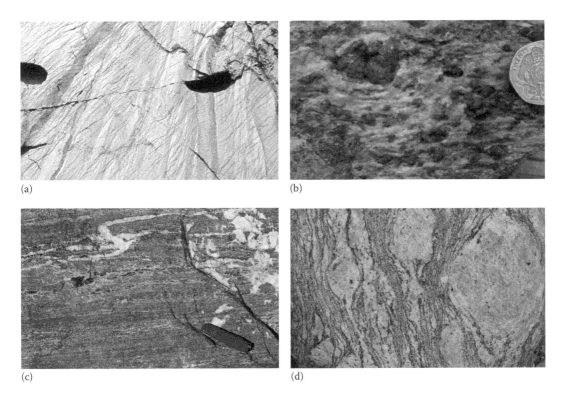

(a)

(b)

(c)

(d)

Figure 7.7 Outcrops of dynamically metamorphosed rocks – original rock type in parentheses. (a) Low-grade phyllite with crenulation foliation covered by shiny white mica (mudstone), Anglesey, Wales. View width 20 cm. (b) Medium-grade garnet (red)- and kyanite (light blue)-bearing schist (feldspathic sandstone). South Harris Scotland. (c) High-grade, amphibole-bearing banded gneiss (volcanic rock), Western Himalaya, Tibet. (d) Augen gneiss (granite), Western Himalaya, Tibet. View width 15 cm.

HYDROTHERMAL ALTERATION

Hydrothermal alteration is the process whereby mineralized hot fluids and gases which are concentrated during the final stages of the cooling of a plutonic magma pass through and alter the previously solidified marginal phase of the pluton and the surrounding rocks. The fluids and gases were channelled along joints, fractures and faults to form a zone of altered rocks rich in kaolinite, chlorite, epidote and quartz, together with rarer minerals such as tourmaline, topaz, sulphide minerals and rarer minerals. Hydrothermal alteration commonly accompanies mineralization and a series of alteration zones, defined by minerals such as sericite (mica) epidote, chlorite and pyrite, are used to help locate mineral deposits. From an engineering perspective, hydrothermal alteration may alter the geotechnical properties of the rock making it either weaker or harder than the original rock. In addition, it may provide an indication of the proximity of faulting that could have acted as conduits for the hydrothermal fluids. Figure 5.14 provides a schematic representation of the upper parts of a granite pluton and shows the expected distribution of hydrothermal alteration with reference to the morphology of the contact of the granite with the country rock. The following sections describe four common types of hydrothermal alteration: chloritization, silicification, greisenization and kaolinitization.

Figure 7.8 Chlorite selvedge adjacent to a series of joints cross-cutting a granite, Discovery Bay, Hong Kong.

Chloritization

Chloritization is a process by which the primary ferromagnesian silicate minerals (e.g. hornblende and biotite) and the feldspars are progressively replaced by chlorite (Fe, Mg, Al hydrated silicate). Chloritization can occur along and adjacent to fractures and joints (Figures 7.8 and 7.9b) or may pervasively invade the whole rock. In the latter case, the texture and mineralogy of the original rock may be completely destroyed by the alteration process leaving only 'ghosts' of the primary minerals. Plagioclase feldspar is commonly the first mineral to be replaced by light green chlorite (Figure 7.9a), whereas in the more intensely altered rocks, dark green chlorite is more common. Figure 7.9 displays a range of chloritized granites from site investigation boreholes in Hong Kong, starting with weak chloritization where the chlorite is restricted to microveinlets (Figure 7.9b), through the development of chlorite-rich segregations, to intense chloritization where the original granitic texture has been completely destroyed (Figure 7.9d).

Chloritization is most intense in the marginal phase to a plutonic igneous intrusion where hydrothermal fluids and gases are concentrated but also extends into the country rock along joints and faults, which provide conduits for these fluids and gases. The presence of chloritization in the country rock may indicate the proximity of a fault or contact with an igneous intrusion.

Silicification

Silicification results from the saturation of a rock by silica-rich hydrothermal fluids, and most, if not all, the constituent minerals are replaced by cryptocrystalline quartz. These hydrothermal fluids have low temperatures and are generated in the later stages of an igneous cycle when either igneous fluids or groundwater become saturated in silica. It is commonly associated with hot springs related to volcanic activity. Faults may become silicified after the main displacements have ceased when silica-rich fluids pass along sheared or brecciated fault rock. Silicification is commonly associated with the development of crystalline quartz veins

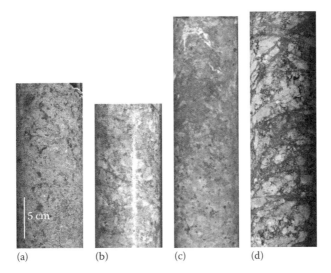

Figure 7.9 Chloritization of granite seen in site investigation boreholes, Hong Kong. (a) Weak chloritization of ferromagnesian minerals, mainly biotite and hornblende. (b) Chloritization restricted to micro-veinlets. (c) Moderate chloritization concentrated within lenticular segregations. (d) Strong chloritization with almost all the plagioclase feldspar, biotite, and hornblende replaced by chlorite – total destruction of granite fabric.

as a result of brittle fracture. Figure 7.10 displays a selection of silicification textures observed in site investigation boreholes from Hong Kong. These show varying intensities of silicification within medium-grained granites and shear zones: first, relatively weak silicification with a network of fine quartz veinlets in granite (Figure 7.10a); second, strong silicification within a shear zone containing deformed lenses of pure quartz lenses within a microcrystalline quartz matrix (Figure 7.10b); and finally, intense silicification with microcrystalline quartz replacements and veins in a brecciated granite (Figure 7.10c).

Silicification may be selective within a fault zone where the pathways for the hydrothermal fluids are controlled by the permeability of the fault rocks. Figure 7.10d displays a

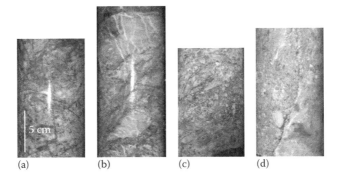

Figure 7.10 Silicification of granite seen in site investigation boreholes from Hong Kong. Diameter of cores 8 cm. (a) Partial silicification of medium-grained granite with network of thin quartz veins. (b) Silicified shear zone with deformed quartz lenses. (c) Grey microcrystalline quartz replacements and veins in brecciated shear zone in granite. (d) Selective hydrothermal alteration of shear zone in granite: the right-hand side of a shear surface has been intensely silicified.

vertical contact between silicified and chloritized fault rocks. The contact is sharp and not displaced but certainly acted as an aquiclude for the silica-rich fluids.

Silicification greatly increases the overall strength of a rock and may anneal part or all of a fault zone that was originally soft and friable. This may be beneficial in that the fault ceases to be a zone of weakness, deep weathering or water ingress, but elsewhere partial silicification may be the result in the formation of quartz lenses in predominantly soft material – a possible problem for tunnel boring machines.

Greisenization

Greisenization is the process by which a rock is replaced by granular quartz and white mica. It is commonly associated with the introduction of boron and fluorine that results in the crystallization of tourmaline (Na, Mg, Fe, B, Al silicate), fluorite, beryl and topaz. Greisens may also contain economic quantities of tin and tungsten. Greisenization forms close to the margins of some plutonic igneous bodies and penetrates the adjacent country rocks where it preferentially alters the rocks adjacent to joints or other discontinuities (Figure 7.11a). In places, greisenization is accompanied by potassic alteration zones, where potassium feldspar has replaced many of the original minerals (Figure 7.11b).

Patches of extremely weak material formed by greisenization may have an effect on the geotechnical characteristics of the rock mass close to the margins of a plutonic igneous body and can lessen the friction angle of joints. In tropical climates, greisen zones weather more readily than the host rock due to their high mica content which readily decomposes to kaolin. In such zones, the rockhead will be depressed in relation to the host rocks.

Kaolinitization

Kaolinitization refers to the pervasive hydrothermal alteration of feldspars within plutonic igneous bodies, in particular granite, to kaolin ($Al_4Si_4O_{10}(OH)_8$). Wide tracts of granite may be kaolinitized, and these are an important source of industrial clay used in the paper and china industries. However, as will be explained in Chapter 11 on weathering, kaolin may

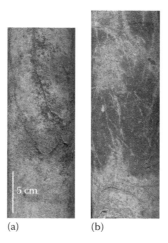

5 cm

(a)   (b)

Figure 7.11 Greisenization of volcanic rocks from site investigation boreholes, Hong Kong. (a) Greisenized medium-grained granite. The alteration follows a joint but also, in places, pervades the rock mass. The geisen zone has a granular texture and is rich in white mica. (b) Network of narrow greisenized veinlets in a crystal-bearing fine ash tuff. The shiny light greenish-brown mica is associated with a zone of pink potassic alteration.

Figure 7.12 Kaolinitization of granite in Hong Kong. The gullied ground is characteristic of this style of intense alteration.

also develop in significant quantities in the tropical soil profiles. The distinction between the two modes of formation is much debated, and it has even been suggested that the origin of the deeply kaolinitized granites of Cornwall, rather than being hydrothermal, could have been, in part, the result of ancient tropical weathering.

Whatever the origin of the alteration, kaolitization intensely weakens the granite leaving it highly susceptible to erosion. The resulting 'badlands' topography is characterized by deep gullies with weak crumbling escarpments (Figure 7.12).

Skarn

Skarn is formed where granite plutons have intruded limestone or calcareous sedimentary rocks, here the country rocks may be strongly replaced by massive concentrations of iron oxides and sulphides, quartz and calcsilicate minerals, which include the minerals epidote and calcium-bearing pyroxene, garnet and amphibole (Figure 13.7a). In places, there is also an enrichment of the economic minerals of copper, lead, zinc and tungsten. Such mineral associations are the result of hydrothermal fluids, rich in particularly iron and silica, emanating from the granite and reacting strongly with the calcite and dolomite of the country rocks. In places, the limestone is only partially replaced, as illustrated in the core sample seen in Figure 7.13b and c where pods of epidote have crystallized in the limestone along the contact with a granite intrusion. However, it is not uncommon that all the calcium minerals of the limestone or dolomite are totally replaced so that the skarn contains no free carbonate minerals.

The presence of skarn mineralization indicates that limestone was once present and therefore will be probably encountered close by. For example, Figure 7.13a shows a banded skarn from site investigation boreholes for the development of a new town in Hong Kong. Here, large limestone bodies were found at the same site during later site investigations and these caused significant geotechnical problems (see Case Study 12.2).

VEINS

Veins are found in a wide variety of geological environments associated with faults, tectonic belts, metamorphic terrains, hydrothermal systems and igneous intrusions. The low-temperature

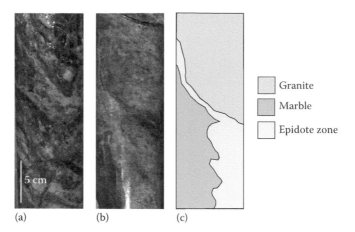

(a)   (b)   (c)

Granite
Marble
Epidote zone

Figure 7.13 Skarn from site investigation boreholes, Hong Kong. (a) Layered skarn with bands rich in epidote, magnetite and amphibole. A few thin layers of the original limestone remain. (b) Yellowish-green skarn developed in grey limestone along the contact with pink granite. (c) Diagram showing the relationships in the adjacent core photograph.

fluids that are generated in these environments permeate through the rock and fill any available open spaces, either in the form of extensional cracks or in shattered rock. Veins are one of the most distinctive features of many rock outcrops as they commonly cut across primary rock structures and many visually stand out due to their high quartz or calcite content. The distribution and form of many veins are controlled by the tectonic stresses active through time after the formation of the rock and therefore can provide sequential evidence for the uplift, faulting and intrusion of igneous bodies.

Veins are important economically as they constitute a major source for the ores of gold, copper and tungsten among many other metals. From a geotechnical engineering standpoint, some veins can define distinct discontinuities within the rock, whereas others, depending on their composition, can form extremely hard masses, both within rock and soil that could adversely affect drilling, pile driving and tunnel boring.

Vein form

Veins can take on many forms ranging from simple planar sheets, *en echelon* sets, folded structures to complex brecciated networks (stockworks) – all of which have been determined by the stress history of the area. They range in width from several metres to less than a millimetre and can have lateral extents of hundreds of metres; however, lengths of few metres or less are the most common. The study of veins is best undertaken at outcrop; however, rock cores can provide vital indications of the vein distribution and the geological history of the project site. The vein form and orientation can be related to the strain ellipse (Figure 7.14) in that the planar veins form at right angles to the maximum extension direction, and the sigmoidal *en echelon* veins are parallel to the conjugate shear directions.

The most veins are part of planar (Figure 7.15a), *en echelon* (Figure 7.15b) and conjugate (Figure 7.15c) arrays. In places, the *en echelon* veins may display sigmoidal shapes associated with shear displacements during the progressive growth of the veins. Irregular vein breccias (Figure 7.15d) are formed as result of fracturing by extreme hydrostatic pressures

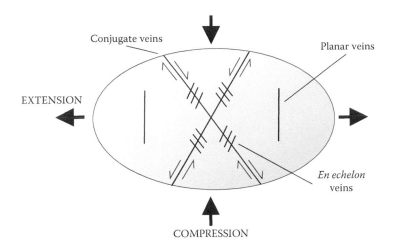

Figure 7.14 Vein arrays in relation to the strain ellipse.

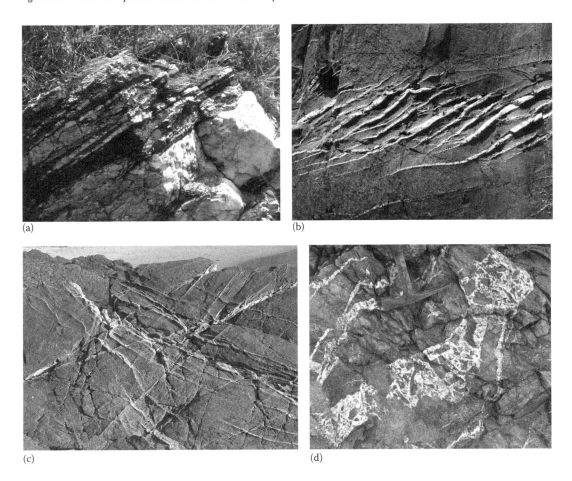

(a)

(b)

(c)

(d)

Figure 7.15 Vein arrays. (a) Variably thick planar quartz veins developed sub-parallel to the foliation in schist, Hong Kong. (b) *En echelon*, slightly sigmoidal quartz veins cutting sandstones, Zambujeira, Portugal. (c) Conjugate quartz vein array in sandstone, Zambujeira, Portugal. (d) Quartz breccia and associated quartz veins in fine-grained altered granite, Anglesey, Wales.

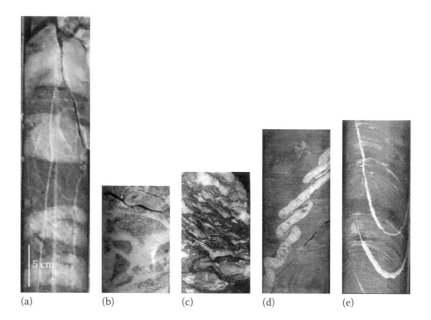

(a)   (b)   (c)   (d)   (e)

Figure 7.16 Vein arrays in borehole core from Hong Kong, except where stated. (a) Parallel, planar veins in volcanic rock. (b) Stockwork of quartz veins within a brecciated zone in altered granite. (c) Deformed and brecciated quartz veins forming part of a gold deposit, Thailand. (d) Asymmetric, isoclinal folds in deformed quartz vein cutting marble. (e) *En echelon* quartz veins in bedded volcanic rock.

Figure 7.17 Multiple vein intrusions with planar, *en echelon* and sigmoidal forms in volcanic rocks, Cadair Idris, Wales.

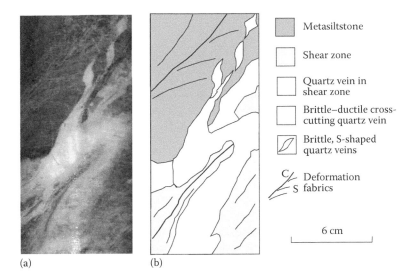

Metasiltstone

Shear zone

Quartz vein in
shear zone

Brittle–ductile cross-
cutting quartz vein

Brittle, S-shaped
quartz veins

Deformation
fabrics

6 cm

(a)   (b)

Figure 7.18 Shear zone with three generations of quartz veining in metasiltstone from site investigation vertical borehole, Tin Shui Wai, Hong Kong. (a) The earliest planar quartz vein is associated with ductile shear deformation in the metasiltstone (S–C fabrics – see Chapter 9); the S-shaped quartz vein with ductile–brittle deformation and the *en echelon*, sigmoidal quartz lenses with the last phase of brittle deformation. (b) The slight greenish tint to the metasiltstone has resulted from low-grade metamorphism of the siltstone. The geotechnical implications of these phenomena are discussed in the text. Tin Shui Wai, Hong Kong.

and may be mineralized (Figure 7.15d). Figure 7.16a–e provides a selection of different vein arrays in core samples (planar, *en echelon*, breccias) from Hong Kong and Thailand, each of which has distinct tectonic significance and possibly geotechnical connotations.

Veins commonly display evidence for multiple phases of formation (Figure 7.17) that can provide a detailed history of the development of regional or local stress patterns over time, even in core samples from site investigation boreholes. For example, an extensional planar vein in marble from Hong Kong (Figure 7.16d) has undergone later ductile deformation and the formation of isoclinal folds, and early quartz veins in a schist from Thailand (Figure 7.16c) have been brecciated and then sheared during later deformational events. Description of veins in core following recognized logging procedures can omit to mention features that are important components of a robust and comprehensive geological model. For example, the core from a site investigation in Hong Kong (Figure 7.18) was simply described in the borehole logs as a 'narrow quartz vein' in siltstone. However, the quartz vein is contained within a narrow shear zone that provides essential information as to the tectonic history of the site. Three generations of quartz emplacement are present: an early planar vein associated with brittle extension, a folded vein that developed during a later brittle–ductile deformation and finally a series of *en echelon*, sigmoidal, quartz lenses that crystallized during the last phase of brittle deformation. The recognition that the quartz vein is part of a multiphase shear zone that was active over an extended period of time would suggest that other shear zones, maybe wider, are likely to be present in the immediate area, and these could contain weak ground or act as aquicludes. With this knowledge not shown in the borehole logs, subsequent ground investigations could be better targeted to locate and characterize any additional shear zones that could be present.

CASE STUDY 7.1 METAMORPHIC ROCKS: MOUNTAIN ROAD CONSTRUCTION, EAST NEPAL

[26°50.2′N 87°16.6′E to 26°59.1′N 87°19.9′E]

R.P. Martin
GeoconsultHK

The 51 km long Dharan–Dhankuta Road traverses the first two ranges of the Low Himalaya of East Nepal and attains a relative relief of c. 1400 m (Figure CS7.1.1). Investigations and reviews for the project included geomorphological surveys for alignment selection (Brunsden et al., 1975); effects of major rainstorms on slope stability, fluvial erosion, flooding and sedimentation (Brunsden et al., 1981); geological and geotechnical input to design and construction (e.g. Fookes and Marsh, 1981; Fookes et al., 1985); post-construction earthquakes (Martin, 2001; Dugar, 2013) and post-construction performance (Hearn, 2002). Approximately 300,000 m³ of gabion retaining walls and revetments were constructed supplemented by about 30,000 m³ of masonry. Over 300 culverts were constructed, with spacing generally <300 m and a peak culvert density of 15 km⁻¹.

GEOLOGY

The geology of the road from south to north comprises an initial small section of unmetamorphosed Tertiary sandstones and siltstones near Dharan; then Permo–Triassic phyllites, amphibolites and quartzites; and finally into Pre-cambrian to Palaeozoic mica schists and gneisses (Figure CS7.1.2). The regional structure is dominated by a series of southwards-directed thrust faults that separate the three main tectonic units. Thick transported soils, especially coarse colluvium and talus mantle most of the steep hillsides, overlying variably weathered rock, except in the gneissic terrain where deep *in situ* rock weathering has created residual soils and saprolites up to 20 m thick. The combination of rapid uplift along the seismically active Himalayan mountain front, steep slopes in fractured and folded rocks and seasonally intense monsoon rainfall means that regional rates of landsliding and gully erosion are extremely high (Brunsden et al., 1981). Geological factors had an enormous influence on the planning, design and construction of the road (Figure CS7.1.3).

Figure CS7.1.1 Hairpins on the Dharan–Dhankuta Road during construction in 1978. The Indus Plain can be seen in the background.

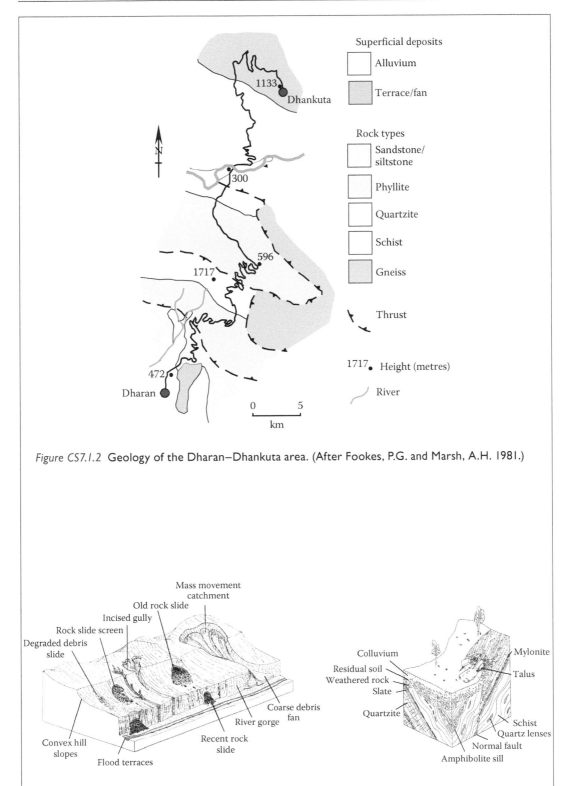

Figure CS7.1.2 Geology of the Dharan–Dhankuta area. (After Fookes, P.G. and Marsh, A.H. 1981.)

Figure CS7.1.3 Schematic terrain models and associated processes. (After Brunsden, D. et al., 1981.)

(a)   (b)

Figure CS7.1.4 Examples of terrain features: (a) Steep quartzite rock faces and talus slopes and (b) soil-covered lower slopes.

For example, where steep rock faces in quartzite and associated talus slopes were encountered (Figure CS7.1.4a), full-cut road cross-sections were preferred, avoiding where possible adversely oriented daylighting discontinuities; in contrast on over-steep, soil-covered, lower slopes, a combination of retained cut and fill was generally chosen in order to minimize overall earthworks and make full use of sound rock at depth (Figure CS7.1.4b). Understanding the evolution of the regional landscape was essential for the design of the road, as Quaternary uplift generally outpaced denudation and valley incision with slope steepness generally increasing towards valley floors. Here, full fill cross-sections with erosion protection below maximum flood levels were generally used (Figure CS7.1.5), but additional works were required where river flow constrictions increase the degree of erosion and scour.

Figure CS7.1.5 Road section constructed across narrow valley floor of a steeply incised river.

Other areas where geological knowledge played an important role in the project included (1) design of offsite drainage and erosion protection works during construction (Fookes et al., 1985), (2) extensive use of bioengineering in the post-construction stage to reduce erosion rates on man-made and natural slopes (Howell, 1999) and (3) design of remedial works for two sections of the road destroyed by the M6.6 Nepal earthquake of 21 August 1988 (Martin, 2001).

ENGINEERING CONSIDERATIONS

- Terrain modelling and near-surface geology characteristics were crucial elements in choosing the final road alignment and appropriate cross-sections.
- Metamorphic fabric and mineralogy had a major impact on the quality and suitability of excavated and crushed rock for construction materials.
- The final alignment was based on rapid ascent by hairpins through relatively stable climbing corridors in the lower and middle slopes and making distance along flatter ridgetops and valley floors.

CASE STUDY 7.2 HIGH-GRADE METAMORPHIC ROCKS: SHAFT AND TUNNEL, LAKE MEAD, UNITED STATES

[36°4.16′N 114°48.08′W]

T.W. Berry, S.E. Pollak and M.L. Piek
Arup New York, Houston and San Francisco, respectively

In 2008, the Southern Nevada Water Authority awarded the Lake Mead Intake No. 3 – Shafts and Tunnel contract to a joint venture between Impregilo SpA and SA Healy, the Vegas Tunnel Constructors. Following the award, the 9 m diameter, 185 m deep access shaft and part of the 20 m wide, 10 m high and 150 m long tunnel boring machine (TBM) launch chamber were successfully constructed. Difficult conditions encountered in the original starter tunnel during 2010 resulted in a new alignment early 2011. Vegas Tunnel Constructors successfully completed the 150 m long realigned starter tunnel and the installation and erection of the 7.22 m diameter 190 m long Herrenknecht AG Dual Mode Mixed Shield TBM, that was needed to construct the 4.7 km long, 6.1 m ID tunnel under Lake Mead, in 2011. The TBM would tunnel under Lake Mead to imbed into the intake structure constructed at the bottom.

GEOLOGY

The project area is located along the western shoreline of Lake Mead in the Boulder Basin (Figure CS7.2.1). The regional geology is dominated by Tertiary sedimentary and volcanic bedrock, as well as Quaternary superficial sediments. Pre-cambrian, metamorphic and igneous rocks are generally confined to the area around the site (after Beard et al., 2007). The realigned starter tunnel was generally constructed in very strong (UCS > 100 MPa), very hard, dark greenish grey, fine to medium grain size, slightly weathered amphibolite and gneiss (Figure CS7.2.2). Joints are generally medium spaced (200–600 mm), persistent to between 0.5 and 3 m, planar to irregular, tight, fracture surfaces are iron stained penetrating ≤2 mm. Faults are very widely

Figure CS7.2.1 Lake Mead with gneiss in the foreground and iron-stained volcanic rocks on the far shore.

(a)   (b)

Figure CS7.2.2 Surface rock exposures close to site. (a) Gneiss. (b) Amphibolite.

spaced (>2000 mm), irregular, open ≤50 mm and filled with (soft) orange-brown clayey sand and fine angular gravel; fault surfaces are iron stained penetrating ≤10 mm. Groundwater flows exceeding 1818 l/min were recorded on a single ungrouted tunnel face during the construction of the original starter tunnel, yet during the excavation of the realigned starter tunnel after pre-treatment by injection grouting reduced volumes of groundwater flows between zero and 181 l/min were recorded per advance.

ROCK MASS CHARACTERIZATION

The three support classes developed for the realigned starter tunnel were linked to the rock mass conditions through the Geological Strength Index (GSI). The GSI is a visual characterization tool developed by Marinos and Hoek (2000), which assigns values between 0 and 100 based on the degree of interlocking between rock blocks and the joint surface (boundary) conditions. The range of GSI values over which a particular support class was considered applicable was determined through the numerical modelling methods using lower bound GSI values to derive the geotechnical parameters used.

ROCK MASS DESCRIPTION

Face mapping was carried out after each round of drill and blast during the construction of the realigned starter tunnel. Each tunnel advanced was between approximately 0.9 and 3.3 m and as such 69 numbers of faces were mapped during the construction phase (Figure CS7.2.3).

The tunnel was mapped (using drawings and photographs) and logged using several classification schemes, including geological classification using BS5930, discontinuity descriptions from Description of Rock Masses for Engineering Purposes and rock mass descriptions using GSI (Marinos and Hoek, 2000), the Rock Mass Rating system reference (Bieniawski, 1989) and the Q system (Barton, 2002).

The rock mass descriptions were used to verify (ground truth) the ground model described in the Geotechnical Baseline Report and define the support class required after each round of blasting. In addition, parameters determined from field observations were used to verify the numerical analyzes used for the design of the support system.

Probe drilling was carried out approximately 50 m ahead of the tunnel. The information gathered by drilling the probe holes captured information about zones of high water inflow and poor rock quality and was used entered into a 3D AutoCAD model (Figure CS7.2.4) and used to characterize the anticipated forward ground conditions to predict behaviour and plan for the required support. A weekly summary of the tunnel face mapping was prepared as part of the Tunnel Design Engineer's weekly report.

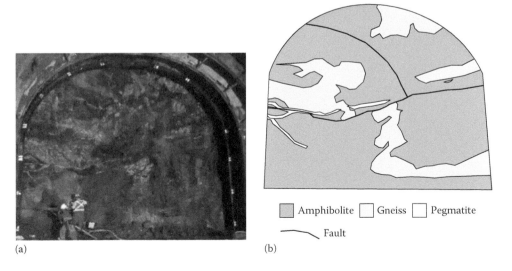

(a)   (b)

Figure CS7.2.3 Typical tunnel face (a) Folded and fragmented, light-coloured gneiss in amphibolite. Pegmatite veins associated with sub-horizontal fault. (b) Main geological features visible on the face.

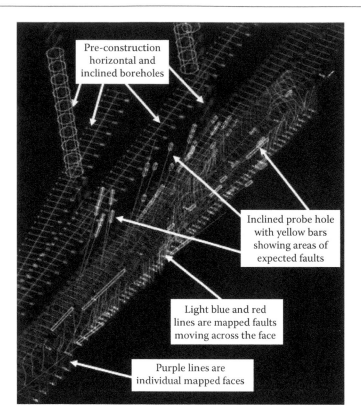

Figure CS7.2.4 3D ground model of the construction data and mapped and expected faults in the tunnel.

ENGINEERING CONSIDERATIONS

- The geologist should see every face, every time, and consistently report the facts to an agreed mapping methodology and presentation format as soon as possible.
- Stronger rocks' performance during excavation is dominated by the discontinuities that need to be mapped accurately and understood completely.
- Groundwater ingress was dominated by certain fracture sets and faults. Understanding the groundwater regime and which discontinuities are likely to produce what quantity of water ingress is paramount to successful underground construction.
- Discontinuity morphology, weathering and infill can vary rapidly and so accurate description and assessment of the significant joint sets is vital.
- The very nature of metamorphic rocks means that their structure, discontinuities and performance can be dominated by the foliations, so these need to be considered in detail to understand their significance for excavation and support strategies.
- Rapid variation in geology needs to be assessed for its impact on the behaviour of the rock, its excavatability and subsequent support.

Chapter 8

Faults

Faults and their associated structures are, in many cases, the most important features to be assessed in many engineering site investigations. They cut, deform and displace the original fabric of the rocks (Figure 8.1a and b), thereby causing them to become, in general, weaker and more fissile, increasing the susceptibility of the ground to weathering and influencing the hydrogeological regimes. Faults range from a few centimetres to several hundreds of metres in width across a fault zone. However, due to the influx of hydrothermal fluids, some fault zones may be re-cemented by quartz, calcite and other secondary minerals (see Case Study 8.1). In these cases, parts of the fault zone can form extremely strong rock masses.

Faults require special attention in many ground engineering projects, and site investigations must be designed to fully investigate the distribution, orientation and composition of any faults. A large proportion of faults are sub-vertical, and therefore site investigations which rely only on vertical boreholes will certainly be inadequate, for they may fail to intersect some faults altogether and not sample the complete range of material within the fault zone. In addition, some fault material may be extremely weak and specialized drilling techniques may be needed to enhance recovery.

This chapter presents a brief overview of the geometry of faults, a description of fault materials to be expected and reviews of the techniques available to investigate faults in the regional context prior to any site investigation. For a fuller description and analysis of faults, the reader is recommended to read Ramsey and Huber (1987) and van der Pluijm and Marshak (2004).

ENGINEERING CONSIDERATIONS

Faults are one of the most important geological features that have to be fully understood in any engineering project. They are particularly relevant to the construction of dams, formation of deep foundations and excavation of subsurface structures such as tunnels and underground caverns. In areas where faults are present, it is essential that a thorough desk study is undertaken to establish the geometry of the fault systems, to determine possible extrapolations of known faults from outside the project area and to investigate the range of fault material to be expected. Initially the desk study should examine published geological maps, satellite imagery and aerial photographs to establish the regional tectonic context. Site investigation studies should initially use site geological surveys and where necessary a range of geophysical techniques to further establish the exact nature and distribution of the known and suspected faults within the site area. The boreholes should be planned to intersect the faults and recover as much material as possible. Inclined boreholes should be used to investigate sub-vertical faults, and specialist drilling techniques utilized to sample the weak material (e.g. fault gouge) that are either flushed out or pulverized during the drilling and recovery process.

(a)

(b)

Figure 8.1 Fault exposures. (a) Fault in cliff section with both lateral and vertical components, Dunraven Bay, Wales. (b) Reverse, sub-vertical fault in a low foreshore exposure that has displaced a bedded sandstone and mudstone sequence, Tolo Channel, Hong Kong.

Engineering consideration should be given to the characteristic features of faults and shear zones that are itemized as follows:

- Fault orientation, geometry and composition can generally be predicted within any fault system.
- Fault arrays within a fault system are consistent down even to outcrop level.
- Although faults can have considerable lateral and vertical continuity, it is not uncommon for faults to terminate or appear within an area.
- The composition of most fault materials is weak unless the matrix and open spaces are replaced or filled by secondary minerals where the fault may become extremely strong and cohesive.
- Brittle faults generally have sharp boundaries whereas there is a gradation of deformation towards ductile faults.
- There may be sharp changes of strength of the fault material within a single fault zone, particularly where the movement has been multiphase.
- A fault zone can act as an aquifer or aquiclude dependent on the composition of fault material, but the water flow may be extremely variable within the fault zone.

Faults occur in many ground engineering projects and were encountered in Case Studies 5.1, 5.2 and 9.1. In this chapter, a multiphase fault zone approximately 20 m wide is described in detail, as it was intersected in a bored tunnel some 150 m below the sea floor. The fault cuts a homogeneous, medium-grained granite and is composed of an interdigitation of fault breccia, foliated rocks and fault gouge and is associated with hydrothermal alteration and the injection of a series of quartz veins.

FAULT GEOMETRY

Fault geometry is determined by regional stress patterns within the Earth's crust that mostly result from the movement of tectonic plates, although other mechanisms such as intrusion of igneous bodies or formation of salt domes can also generate faults. There are

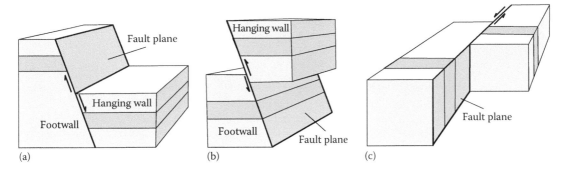

Figure 8.2 Diagrams showing relative movement of blocks across (a) normal, (b) thrust (reverse) and (c) strike-slip (wrench) faults. The sense of movement in the strike-skip fault illustrated is described as left lateral; if the movement had been in the other direction, it is called right-lateral. See Figure 5.16 for right-lateral displacement of the Great Dyke of Zimbabwe. The relative positions of the hanging walls and footwalls of normal and reverse faults are indicated.

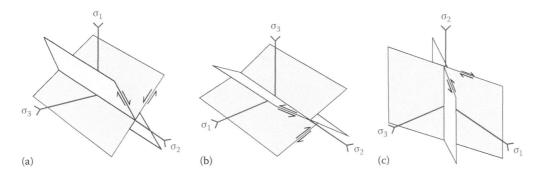

Figure 8.3 Orientation of (a) normal, (b) thrust and (c) strike-slip faults with respect to the principal stresses: σ_1, maximum stress; σ_2, intermediate stress; and σ_3, minimum stress.

three main types of fault: normal faults, reverse (thrust) faults and strike-slip faults, which are defined by the relative movement of the rocks on either side of the fault (Figure 8.2). Thrust faults are reverse faults where the dip of the fault plane is low to moderate. The development of the different fault types is controlled by the relative orientations of the regional stress patterns (Figure 8.3). Normal faults form perpendicular to the maximum extension direction, thrust faults perpendicular to the maximum compressive direction, and strike-slip faults at acute angles to the maximum compressive direction. Oblique displacements on faults are the result of a combination of the fault types.

Normal faults

Normal faults form in response to extension as a result of pulling apart of the Earth's crust. Groups of normal faults commonly define the margins of sedimentary basins or rift valleys, for instance the East African Rift Valley which can be traced over many hundreds of

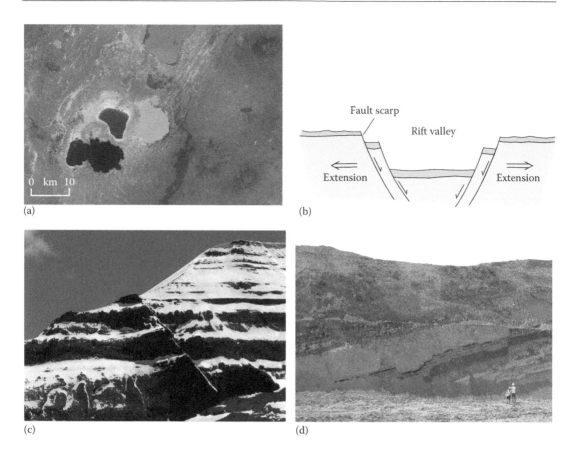

Figure 8.4 Normal faults. (a) Google Earth, Landsat image of the East African Rift Valley. A series of topographic fault scarps define the walls of the rift, Lake Zway. Ethiopia [7°29′N 38°32′E]. (b) Schematic cross-section of a rift valley flanked by normal faults. (c) Planar, normal fault cutting thickly bedded, quaternary sandstones and conglomerates, Mount Kailash, Tibet. (d) Antithetic normal faults defining a small graben (rift) structure. The fault surfaces steepen upwards, western Turkey.

kilometres (Figure 8.4a and b). They are generally steeply dipping (Figure 8.4c) and contain fault breccia or gouge formed by the fragmentation and grinding up of the adjacent rocks. In rift valleys or down faulted blocks, the dip of the fault planes oppose one another and tend to become steeper towards the ground surface (Figure 8.4d).

Thrust faults

Thrust faults generally form in fold belts formed by shortening of the crust during mountain building episodes related to the collision of tectonic plates. For example, in the Zagros Mountains of Iran, thrusts are associated with a series of folds whose axes are orientated parallel to the strike of the thrust faults (Figure 8.5a and c). The thrust faults have low to moderate dips (Figure 8.5c) and can vary in thickness from a single narrow surface to wider zones of shearing. Commonly, thrust faults form a stack of overlapping rock slices above a main subhorizontal detachment thrust (Figure 8.5d).

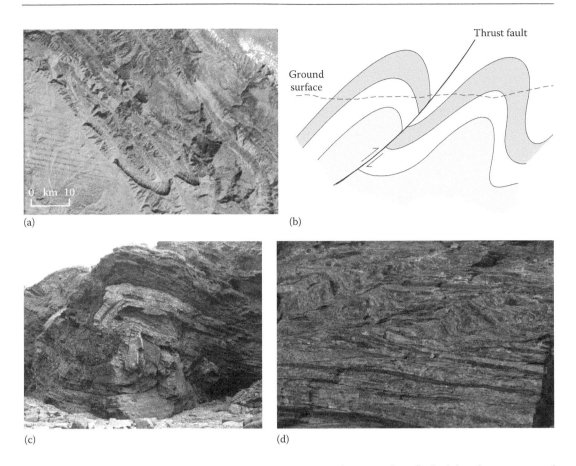

(a)

(b)

Ground
surface

Thrust fault

(c)

(d)

Figure 8.5 Thrust faults. (a) Google Earth, Landsat image of an anticline flanked by thrusts, central Iran [32°30′N 49°13′E]. (b) Schematic cross-section across a thrust and associated folds. (c) Low-dipping thrust fault with associated anticline and syncline in interbedded sequence of Carboniferous shale and sandstone, Broadhaven, Wales. (d) Series of small-scale thrust faults in quartz schist, forming an imbricate stack bounded below by a basal, near-horizontal thrust, Cameron Highlands, Malaysia. View width 15 cm.

Strike-slip faults

Strike-slip faults commonly develop during the oblique collision of tectonic plates that results in horizontal displacement of sections of the crust. This displacement is described as either left lateral as in Figure 8.2 or right lateral if there is the opposite sense of movement. For example, along the boundary of the northward-moving Indian Plate and the Afghanistan Plate (Figure 8.6a and b), there has been hundreds of kilometres of left-lateral strike-slip displacement along the Chaman Fault since the initial collision of the two plates. This fault is still active today, and recent displacements of drainage courses can be seen on the satellite images when viewed in close-up. Strike-slip faults generally have steep dips and are commonly associated with zones of shearing, folding and anastomosing (braided) fault arrays (Figures 8.6c and 8.7).

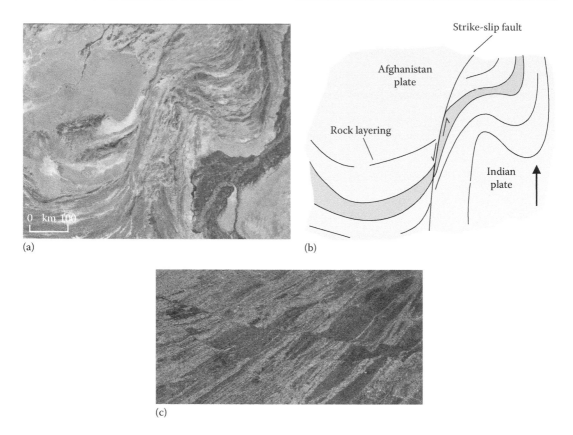

(a)

(b)

(c)

Figure 8.6 Strike-slip faults. (a) Google Earth, Landsat image of the Chaman Fault, Pakistan–Afghanistan border region [30°20′N 66°20′E]. Displacement along the fault is over 200 km, due to the northward movement of the Indian Plate. The fault is still active, and the recent movement may be seen, when the image is magnified, by displaced drainage lines across the fault. (b) Schematic sketch of features related to the Chaman Fault. (c) Strike-slip fault with lenticular branches cutting vertical Archean banded gneisses, Kerala, India. 2.5 cm coin for scale.

FAULT SYSTEMS, ARRAYS AND ZONES

A group of faults that formed under a consistent regional stress regime and over a particular time frame is said to constitute a *fault system*. The distribution of the faults with a system may include different fault arrays that define the geometry of the faults. These are classified into parallel, anastomosing, *en echelon* and conjugate arrays (Figure 8.7a). Parallel fault arrays are made up of series of subparallel faults, anatomizing fault arrays comprise braided series of connecting faults, *en echelon* fault arrays are non-continuous faults that occur in stepped segments, and conjugate fault arrays are formed of two series of faults that intersect at about 60° to each other.

Fault systems can dominate the geological development of a region; for instance in eastern China where the north-east-trending fault system can be traced for over 2000 km – it has largely controlled the tectonic, sedimentary, volcanic and intrusive history of the region since the Mesozoic times and contains a variety of fault arrays (Figure 8.7b). It is very important when assessing a particular site for engineering purposes to establish whether the site lies within a particular fault system in order to anticipate and interpret the geology correctly. Geological development and geometry of a fault system as seen on regional

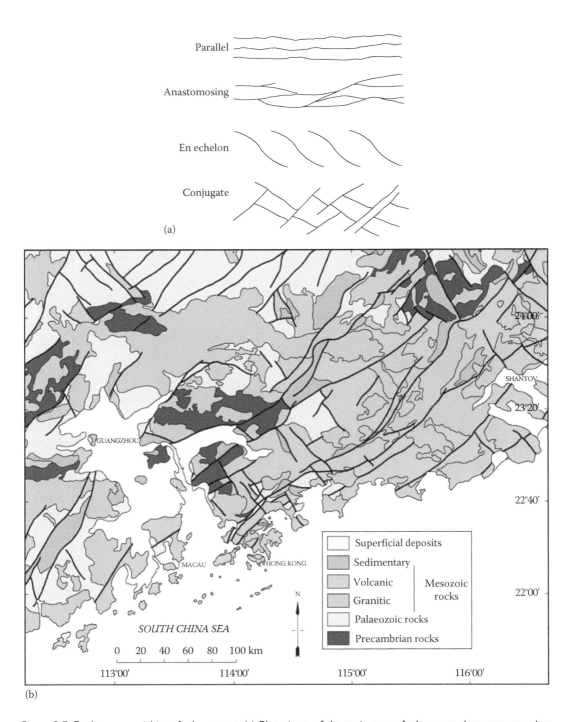

Figure 8.7 Fault arrays within a fault system. (a) Plan views of the main types fault arrays that accommodate the overall displacement across a fault zone. (b) Fault system of south-eastern China composed of parallel, anastomosing and conjugate fault arrays. (After Bureau of Geology and Mineral Resources of Guangdong Province, *Regional Geology of Gungdong Province*, Geological Memoirs Series, Vol. I, 1988, 941pp.)

geological maps are commonly reproduced on a smaller scale and may be recognized at site or even outcrop level. For example, on the regional geological map of the southern part of the southeast China fault system, only two faults are shown that pass through Hong Kong (Figure 8.7b). However, in detail many more faults have been mapped in Hong Kong, and these conform to the regional fault array patterns and geometries.

The intensity of faulting within a fault system may vary considerably, but where there is a concentration of subparallel faults across a zone of definable width, the term *fault zone* is used. Fault zones having regional significance may be tens of kilometres wide, but the term can also be applied to fault zones only tens or hundreds of metres wide. Where the fault displacement is dominated by ductile deformation (see Ductile Fault section), the term *shear zone* is used.

All faults eventually terminate because either the regional stresses may diminish, they pass into a fan of less significant faults, they are truncated by later intrusions and faults, or the displacements are transfered to other subparallel faults. This is illustrated on the China geological map (Figure 8.7b) where in certain localities north-east-trending faults terminate against granite intrusions or north-west-trending faults. Again it is vital to understand the fault distribution patterns within a fault system and therefore be fully aware and anticipate that unmapped faults may be present in the site area.

FAULT COMPOSITION

The composition of a fault is dependent on the mechanism of displacement, the composition of the country rocks, movement history of the fault and degree of secondary mineralization and/or alteration. As a consequence, fault material can vary from extremely soft and fissile material to very hard fault rock, and these may be juxtaposed across a fault zone – a reflection of changing dynamics and hydrothermal conditions during the displacement history of the fault, which may extend over many millions of years. The thicknesses of individual faults can be extremely variable from less than a centimetre (Figure 8.5c), to a few metres (Figure 8.1a), to several tens metres wide (Figure CS8.1.4). Overall it is common for faults to define negative topographic lineaments at surface due to the preferential weathering of the weaker and more fractured fault material and a decrease in joint spacing towards the fault (Figure 9.10a).

The two main mechanisms for the formation of fault rock are brittle failure that occurs at higher levels in the crust and ductile shearing that is more characteristic of deeper levels with elevated temperatures and pressures or where strain rates are low. Idealized block diagrams of brittle to ductile fault displacements are shown in Figure 8.8; however, there is a gradation between the two extremes of displacement from brittle through brittle–ductile

Figure 8.8 Idealized geometries of a right-lateral fault that has been displaced during brittle to ductile conditions. Intermediate conditions display both ductile and brittle geometries.

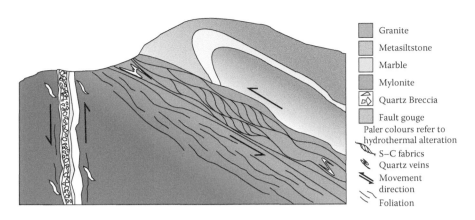

Figure 8.9 Schematic diagram illustrating the components of a normal brittle fault and a ductile thrust fault.

and ductile–brittle to ductile. In general, normal faults develop by brittle failure, whereas ductile shearing is more characteristic of reverse and strike-slip faults. However, a fault may change its character through time as a result of factors such as the depth at which the displacement took place (see 'Multiphase faults' section), change in the regional stress vectors or the duration of the fault movement – for example ductile faults are generally associated with prolonged deep displacements, whereas shallow short-duration movements are more characteristic of brittle structures. A schematic representation of the characteristic fault rocks, veins and foliations to be expected in a typical brittle normal and ductile thrust fault zones is presented in Figure 8.9.

Brittle faults

Displacement along brittle faults occurs mainly by fracturing, fragmentation, grinding and frictional sliding which produces a wide variety of fault rock types, each of which has distinct geotechnical properties. In rare cases, frictional heating caused by the brittle fault movement results in the melting of the rock and the formation of *cataclasite* – a glassy material that can intrude into the adjacent rocks. The brittle deformation can extend beyond the zone of movement through stress transfer and high fluid pressures.

Fault breccias form by fragmentation and granulation of a rock through brittle failure caused by tectonic displacements. Angular blocks of country rock, in places partially rounded by later fault movements, are embedded in a fine-grained matrix (Figures 8.10b and 8.11a). Where the rocks within the fault zone have been severely crushed and ground down to a size of less than 1 mm, the material is referred to as *fault gouge*. Fault gouge varies from angular, sand-sized, mineral and rock fragments set in a silt-sized matrix (Figure 8.11b and c) to a pure clay-rich gouge (Figure 8.11d). Fault gouges commonly have very sharp contacts with the host rock (Figure 8.10a) which results in abrupt changes in rock strength. In some fault zones, renewed movement along the fault may impart slip surfaces or a foliation to the gouge.

Fault gouge can be one of the most important geological features in a fault zone from an engineering perspective, as it is composed of extremely weak non-cohesive material. For example, a narrow seam of fault gouge within a 30 m wide fault zone intersected in a tunnel beneath the sea floor in Hong Kong caused significant delays in the advance of the

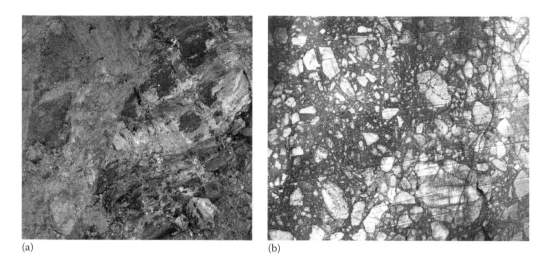

(a)   (b)

Figure 8.10 Brittle faults. (a) Soft, clay-rich fault gouge with angular rock fragments in sharp contact against low-dipping schists in a cut slope, Po Selim, Malaysia. (b) Extremely hard fault rock with unorientated, angular fragments of meta-sedimentary rock set in a very fine-grained, siliceous matrix. View width 1.5 m, northern Pyrenees, France. (Photograph courtesy of W.R. Fitches.)

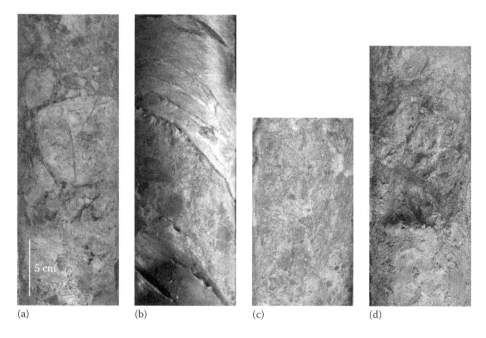

(a)   (b)   (c)   (d)

Figure 8.11 Examples of brittle fault material in the core from Hong Kong. (a) Fault breccia in granite with subangular to subrounded blocks of granite and quartz in a matrix of granulated mineral and rock fragments. (b) Narrow, sharply defined clay-rich fault breccia in marble with angular marble rock fragments in clay matrix. (c) Weak fault gouge in granite with small angular fragments of pink feldspar and quartz set in a granular, light green matrix composed of clay and very fine-grained rock and mineral fragments. (d) Extremely weak, clay-rich fault gouge in calcareous siltstone; the purple iron staining is due to later oxidation.

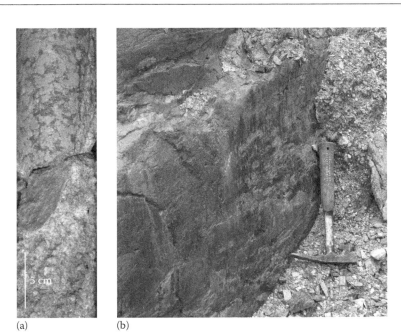

(a)   (b)

Figure 8.12 Slickensides. (a) Slickensided faulted contact between chloritized granite and fault gouge from an inclined borehole, Hong Kong. (b) Slickensides on exposed fault surface (Case Study 9.1) with strong vertical striations and a weak, earlier sub-horizontal striation, Po Selim, Malaysia.

tunnel boring machine (see Case Study 8.1). Fault gouge may have a significant effect on the hydrogeology of the site – granular fault gouge may be a channel for groundwater, whereas clay-rich fault gouges can act as aquicludes. Fault gouge is difficult to recover using conventional drilling techniques for ground investigation, and this may severely jeopardize the geotechnical assessment of the site. The use of foam or polymer flushing fluids for drilling will normally enhance the recovery of fault gouge or other weak fault materials.

The movement of two rock masses against each other can produce a highly polished surface due to frictional sliding, which is referred to as a *slickenside*. Grooves or mineral fibres (quartz, calcite, chlorite), which provide an indication of the direction of fault displacement, can develop on the slickenside surface (Figure 8.12a), and it is not uncommon to be able to identify two or more successive fault displacements (Figure 8.12b).

The passage of aqueous fluids (groundwater and/or hydrothermal fluids) along many faults, subsequent to their initial formation, fills the open spaces between the fragments with mineral precipitates. Fault breccias are commonly cemented by quartz or calcite (Figure 8.13a and b) which may significantly increase the strength of the original fault material and affect the overall hydrological pathways within and adjacent to the fault. Where hydrothermal fluids have passed through the fault zone, the adjacent rocks along the fault are commonly altered and zones rich in secondary minerals such as chlorite and white mica may develop.

Ductile faults

Displacements across ductile faults and shear zones are progressive and no brittle failure has taken place (Figure 8.14b). Ductile faults display a variety of structures, including shear

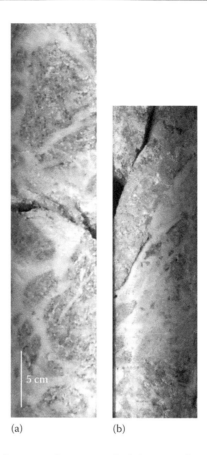

(a)   (b)

Figure 8.13 Cementation and replacement features in fault breccias from Stonecutters Bridge site investigation inclined boreholes (see Case Study 5.1). (a) Fault breccia in volcanic rock with quartz
cement. (b) Calcite cement of granite fragments in fault breccia – some fragments have been
partially to totally replaced by pale green mica and calcite (ghost structures).

foliations (Figures 8.14a and 8.15b), stretched pebbles and intensely folded and deformed
rock layers or veins (Figure 8.15c). Some or all of these can indicate the sense of movement
across the fault zone. The intensity and style of deformation within a ductile fault is dependent on the original composition of the rock, amount of displacement, depth of faulting and
the duration of fault movement.

Shear foliations in ductile faults have distinctive geometries with lenses of deformed rock
being enveloped by the original foliation prior to the ductile shearing (S fabric) and the shear
foliation (C fabric). The angular relationship between the S and C fabrics provide evidence
for the sense of movement along the shear zone (Figure 8.16). In deformed homogeneous
rocks, such as granite, less well-developed S–C fabrics can also develop due to the shear
foliations being formed in two or more stages (Figure 8.15b).

Within ductile fault zones, the original components of the rocks, such as pebbles in conglomerates (Figure 8.15a) or mineral grains in a granite, become deformed. The degree
of deformation of the pebbles or mineral grains is dependent on their hardness, the softer
marble pebbles seen in a core section of a conglomerate display extreme flattening, whereas

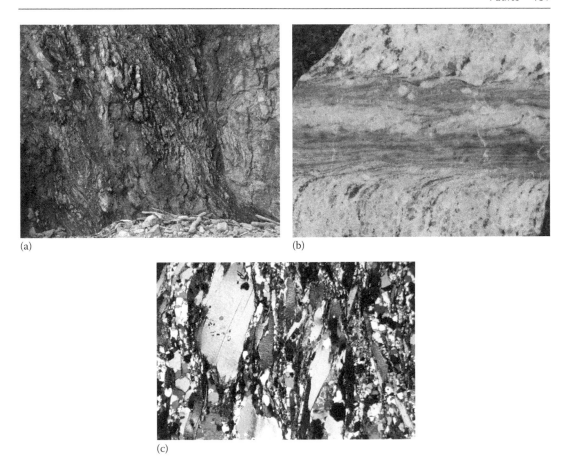

(a)

(b)

(c)

Figure 8.14 Ductile fault zones. (a) Detail of fault shown in Figure 8.1a composed of very fissile mudstone with small fragments of limestone. Fault contact with massive limestone is vertical and
sharp – Dunraven Bay, South Wales. (b) Narrow ductile, mylonitic shear zone in Precambrian
granite – Malawi. The foliation in the granite is bent in opposite directions on either side of the
shear zone. An anastomosing foliation has developed along the margins of the mylonite. The displacement on the fault is right-lateral. View width 10 cm. (Photograph by W R Fitches.) (c) Thin
section of mylonite in ductile fault zone with distinctive wavy foliation, stringers of strained
quartz granules and deformed mica flakes (yellow and blue interference colours). Crossed
polarized light, view width 5 mm – Scotland. (Courtesy of the British Geological Survey.)

the hard quartzite pebbles largely retain their original shapes. In this section, a mylonite can
display a variety of ductile structures including wavy foliation, stringers of strained quartz
and deformed mica flakes (Figure 8.14c).

Original compositional layering or planar veins become intensely folded and deformed in
shear zones, and commonly to such an extent that only the fold hinges remain within the
shear foliation (Figure 8.15c), the fold limbs having been completely sheared out. However,
it is these structures that provide vital evidence for the style of the displacement taking place
within an area.

Where the ductile deformation is extreme, a rock called *mylonite* is formed that is fine
grained and very fissile, with the constituent grains having been deformed plastically

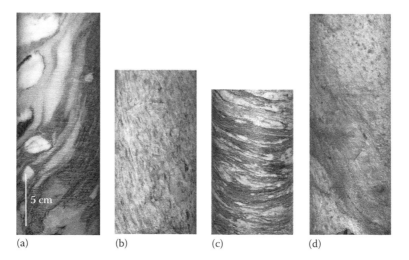

(a)   (b)   (c)   (d)

Figure 8.15 Examples of ductile fault material in the core from Hong Kong site investigations. (a) Intensely flattened pink marble pebbles in contrast with less deformed quartzite pebbles in foliated matrix. (b) Pervasively foliated granite with anastomosing foliation defined by mica and elongated feldspar grains. (c) Ductile deformation of an originally thinly bedded sequence of light grey marble and calcareous metasiltstone. The marble layers have been isoclinally folded and the limbs sheared out. (d) Mylonite in medium-grained granite consisting of very fissile material and narrow bands of recrystallized quartz.

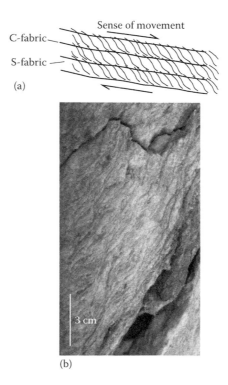

(b)

Figure 8.16 Development of S–C fabrics in shear zones, Hong Kong. (a) Schematic diagram of the relationships between the C and S fabrics that provide the sense of movement within a shear zone. (b) S–C fabrics in a deformed granite within a shear zone.

(Figure 8.15d). In some mylonites, there is a segregation of minerals into discrete bands and stringers, particularly quartz and feldspar, and the growth of 'tails' to rotated minerals.

The margins of ductile faults are not generally sharp, and the deformation extends into the adjacent rock, either by bending of existing foliations (Figure 8.14b) or deformation, such as flattening of pebbles, becoming gradationally more intense as the ductile fault is approached. Such extended influence of a ductile fault zone provides evidence for the proximity of such a fault, even if the fault itself is not exposed or not intersected in boreholes.

Multiphase faults

Faults may display a variety of deformation styles related to successive displacement episodes, which occurred under different stress conditions. For example, a mylonite layer that formed under ductile conditions may be cut by fault gouge that was developed during a later period of brittle displacement. Figure 8.17 shows an example of a multiphase fault in a core sample that developed first as a ductile shear zone in granite, followed by quartz veining related to extension and finally as a brittle fault with the formation of fault gouge. The multiphase fault described in Case Study 8.1 shows evidence of at least three periods of displacement, hydrothermal alteration and quartz veining related to both brittle and ductile deformation. Such reactivation of faults over a considerable time period is not uncommon, because once a fault has been created, it continues to act as a zone of weakness within the crust that is exploited during subsequent deformation events.

Figure 8.17 Multiphase fault zone in vertical rock core, Deep Bay, Hong Kong. Early ductile shear displacement (foliated granite), followed by extensional faulting (quartz vein) and finally late brittle faulting (fault gouge).

CASE STUDY 8.1   FAULT ZONE IN GRANITE: TUNNEL, RAMBLER CHANNEL, HONG KONG

[22°20.4′N 114°7.1′E]

A 3.2 m diameter tunnel was driven by an unshielded boring machine from Tsing Yi to Stonecutters Island (Figure CS8.1.1) as part of an integrated sewage strategy disposal scheme for Hong Kong. The tunnel, referred to as Tunnel F, crossed beneath the Rambler Channel, whose geology at that time was only known from a series of widely spaced vertical boreholes along the tunnel alignment and extrapolations of the rocks mapped on land (Figure CS8.1.2). At approximately 750 m from the access shaft of Tsing Yi, the tunnel boring machine encountered seams of extremely weak material that caused partial collapse of the roof and delayed progress of the work by several weeks. Subsequent geological investigations revealed that the weak material was part of a complex fault zone cutting the homogeneous granite (Sewell et al., 2000).

GEOLOGY

The desk studies and site investigations prior to the start of the tunnel boring suggested that the alignment would pass through fresh medium-grained granite (Figures CS8.1.2 and 8.1.3a) for most of its length and that weathered rock would not be encountered at the depth of the tunnel (135 m below PD). Two north-east-trending faults had been inferred to cross the alignment; however, the fault zone that was intersected was orientated north–south suggesting that a previously unmapped fault was aligned along the Rambler Channel. The width of the fault zone was 15 m and developed during a series of ductile and brittle tectonic episodes. Several sections of fault zone contained very poor quality rock, and high groundwater inflow rates occurred adjacent to the bands of fault gouge and along fault planes. Importantly, the component parts of the fault had widely different geotechnical and hydrogeological characteristics, and therefore each part had to be assessed individually. The fault displays at least three separate phases of displacement: a ductile phase when the granite was sheared, foliated (S–C fabrics) and hydrothermally altered; a second brittle phase associated with quartz veining, brecciation and development of stockworks; and a final brittle phase when the fault gouge was formed.

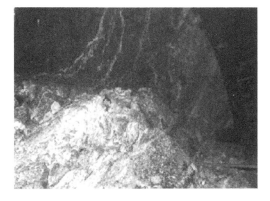

Figure CS8.1.1 Fault rocks exposed in the face of Tunnel F during construction showing a band of quartz breccia within sheared and altered granite.

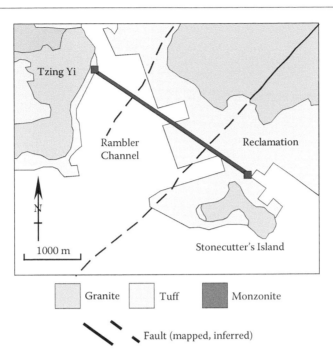

Figure CS8.1.2 Geology of Rambler Channel prior to excavation of Tunnel F (red line). (After Geotechnical Engineering Office, Geological Map of Hong Kong, Millennium Edition, 1:200,000 scale, Civil Engineering Department, Hong Kong SAR Government, 2000.)

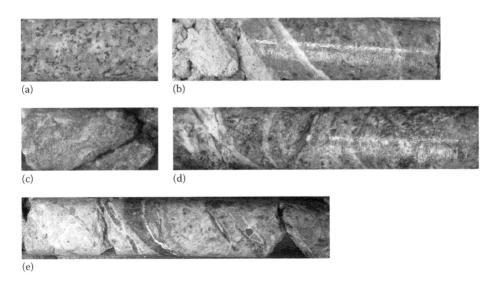

Figure CS8.1.3 Examples of core from a horizontal borehole across the fault zone in Tunnel F. All core diameters 8 cm. (a) Fresh medium-grained granite adjacent to the fault zone. (b) Sharp contact between fault gouge and altered granite. (c) Chloritized and sheared granite displaying pronounced foliation. (d) Stockwork of irregular quartz veins within chloritized granite displaying little of the original granite textures or mineralogy. (e) Fault breccia composed of altered granite cemented by veins and segregations of grey and white quartz.

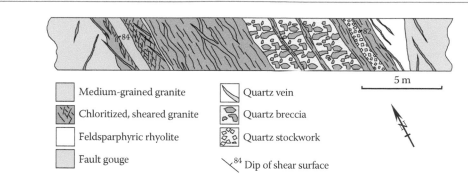

Figure CS8.1.4 Schematic diagram of the fault rocks exposed in Tunnel F. (After Sewell, R.J. et al., 2000.)

The sub-vertical fault zone consists of an interdigitation of fault gouge, hydrothermally altered and sheared granite, mylonite, quartz veins, breccia and stockwork (Figure CS8.1.4). A series of *en echelon* quartz lenses and narrow planar quartz veins occur within the fresh granite close to the margins of the fault zone. These become more prevalent as the main fault margin is approached. Within the fault zone, the granite has been hydrothermally chloritized and sheared to varying intensities. In places, the original fabric of the fresh granite has been totally destroyed by shearing and hydrothermal alteration (Figure CS8.1.3c). The quartz breccia consists of angular altered granite fragments cemented by quartz (Figure CS8.1.3e), whereas the quartz stockwork is composed of an irregular network of narrow quartz veins within altered granite (Figure CS8.1.3d). A 2 m wide very weak fault gouge with very sharp contacts was encountered in the 2 m thick seam (Figure CS8.1.3b). Several distinct fault planes, which show evidence for both vertical and horizontal movements, are present throughout the fault zone.

ENGINEERING CONSIDERATIONS

- The fault zone is associated with quartz veining, brecciation, hydrothermal alteration and shearing related to at least three phases of movement.
- Fault zone consists of an interlayered sequence of very weak and strong materials.
- Fault gouges are less than 2 m thick and constitute no more than 10% of the fault zone.
- Water ingress is greatest along the margins of the gouge and fault planes.

Chapter 9

Discontinuities

The definition of a *discontinuity* in this book is taken to be a surface in a rock mass that has the potential to open during engineering works and has lower strength than the surrounding rock (British Standards Institution, 2003). Examples of discontinuities in rock using this definition are bedding or compositional layering, joints and foliations, together with faults and intrusive contacts which are discussed in separate chapters. However, it is recognized that for engineering purposes, it may be more appropriate to restrict the usage of the term discontinuity to where mechanical fracture has taken place (International Society for Rock Mechanics, 1978; Norbury, 2010; Hencher, 2012), which would only include open fractures such as joints. Whatever the definition one uses, it is the recognition, description and testing of discontinuities that are paramount in any rock engineering investigation.

Discontinuities may form at different stages in the formational history of a rock, from primary depositional layers in sedimentary or volcanic rocks, cooling joints in plutonic igneous rocks, foliations in metamorphic rocks, to joints related to later earth movements. In addition, foliations and joints may themselves exhibit multiple stages of formation. For example, the joints displayed in a homogeneous granite from Hong Kong (Figure 9.1) were formed in two distinct phases – the early sub-vertical joints formed during the regional tectonic evolution of the area, whereas the low-dipping joints formed much later during the erosion and exhumation of the granite in relatively recent times.

ENGINEERING CONSIDERATIONS

Discontinuities are one of the most important features to be assessed in any engineering appraisal of a project site, as they can largely determine the geotechnical characteristics of the rock mass (Hencher, 1987). They also strongly control the weathering characteristics of the rocks and therefore may exert a major influence on the stability of hill slopes and the topographic development of the land surface. This is well illustrated on the large scale by reference to and the study of satellite images, where discontinuities commonly are the most recognizable features of the terrain. For example, the horizontal sedimentary layers along the incised Colorado River (Figure 9.2a) are sculptured into a series of benches formed by the hardest rock layers; a network of intersecting joints within a granite body exposed along the sides of Yosemite Valley, California (Figure 9.2b), have been preferentially weathered out and have controlled the drainage patterns; and the dominant surface feature of the metamorphic rocks exposed across the desert near Aswan, Egypt, is the gneissic foliation (Figure 9.2c). Such features observed at the large scale also have their

Figure 9.1 Joints in homogeneous granite. Sub-vertical joints are tectonic, whereas the low-dipping joints are sheeting joints related to stress relief during the progressive uncovering of the outcrop by erosion of the overlying rock, Po Toi Islands, Hong Kong.

counterparts at project, outcrop and even hand specimen levels, and therefore the study of the easily viewed satellite images can be an essential starting point of any ground engineering investigation.

In many cases, there may be several discontinuities within a particular rock mass, each of which may have different geotechnical parameters such as strength, hydrogeology, fracture potential and weathering characteristics. It is not uncommon that one discontinuity will be the most important feature to be assessed, although there may be several others that greatly influence the appearance of the rock mass. A good illustration of this is the complex geology displayed in the Precambrian rocks from Bangalore, India (Figure 9.3a), which includes compositional layering, gneissic foliation, intrusive granite contacts and veining within a single outcrop. These features visually dominate the appearance of the rock, but it is the regularly spaced joints that constitute the main discontinuity with respect to the geotechnical characteristics of the rock mass. These have determined the fracture potential of the rock to such an extent that the rock readily splits into large slabs (Figure 9.3b). The earlier discontinuities have been fused together during metamorphism and magma intrusion so that the rock acts as a coherent entity and the joints cut cleanly across all preexisting geological features.

Depending on the engineering project, discontinuities have to be assessed as to their form, roughness, spacing, openness, friction angles and infill material. All these parameters are described using conventional notations as presented, for example in British Standards Institution (1999) and Norbury (2010).

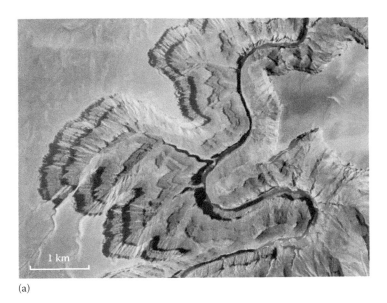

(a)

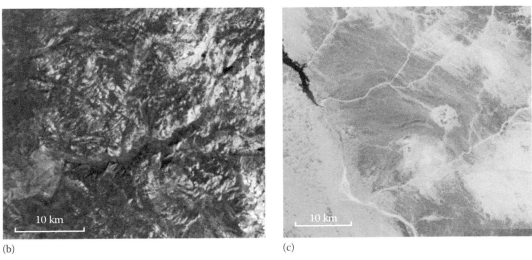

(b)   (c)

Figure 9.2 Google Earth, Landsat images of discontinuities. (a) Horizontal beds in a sedimentary sequence exposed along the Colorado River, United States [36°24′N 111°52′W]. (b) Joint sets cutting granite over the high ground above Yosemite Valley, United States [37°45′N 119°38′W]. (c) Foliation in gneisses, Aswan, Egypt [22°47′N 33°20′E].

The major features of discontinuities that should be understood, described and considered in any engineering investigation are the following:

- Compositional layering, in particular bedding, can result in rocks with extremely different geotechnical properties, hydrogeological regimes and weathering characteristics being juxtaposed.
- Joints are pervasive weaknesses in a rock mass. They have variable strengths dependent on mineralization, infill material and intensity of weathering and can strongly control the flow of ground water through a rock mass.
- Foliations generally weaken the rock mass and can form planes of rupture.

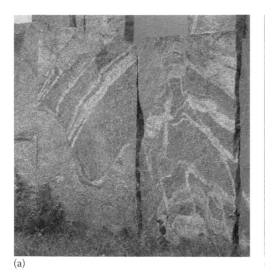

(a)　　　　　　　　　　　　　(b)

Figure 9.3 Regularly spaced joints in a complexly deformed, brecciated and veined gneiss, Bangalore, India. (a) Gneiss slabs in a boundary wall. The bounding joints cross-cut all the previous structures and define the principal planar weakness in the rock; the compositional layering and intrusive contacts have been 'welded' together. (b) Stacked joint-bounded slabs up to 2 m in length with an average thickness of 4 cm.

Case Study 9.1 describes the investigation into a major landslide in a mountainous region of Malaysia, which was initiated by the construction of a road. The geology of the area is dominated by low- to medium-grade metamorphosed mudstone and sandstone that have been highly sheared, folded and faulted. To ascertain the mechanisms that triggered the landslide, a wide variety of discontinuities had to be assessed, including compositional layering, foliations, normal and thrust faults and several sets of joints. The characteristics, distribution and orientations of all these discontinuities had to be thoroughly documented and analyzed in order to postulate on the most probable mechanism for the slope instability. Case Study 9.2 describes the excavation and stabilization of rock cuts up to 15 m high formed during the widening of a road in Wales. Here, the variable orientations of bedding and cleavage within a folded sequence of sandstone and mudstone presented potential for different types of slope failure.

COMPOSITIONAL LAYERING

Layers of different composition occur most commonly in sedimentary and metamorphic rocks and more rarely in igneous rocks. Such layering may result in abrupt changes in rock strength, joint development and weathering intensity and may define significant discontinuities that will affect the geotechnical assessment of a site.

Bedding is the primary stratification seen in many sedimentary sequences and results from a change in the depositional environment, for example the interlayering of mudstone, originally deposited as clay-sized particles from the water column, and sandstone, derived from the influx of coarser sand grains from nearshore environments (Figure 9.4a). The original orientation of bedding is most commonly nearly horizontal as the sediment

(b)

(a)

(c)

Figure 9.4 Varieties of compositional layering in rock. (a) Sedimentary layering in a sequence of alternating beds of sandstone and mudstone, Aberystwyth, mid-Wales. (b) Gneissic layering in high-grade metamorphic rock with bands rich in light-coloured minerals (quartz and feldspar) and bands rich in dark-coloured minerals (biotite, hornblende), Kerala, India. (c) Igneous layering in a pluton where crystals of different composition have settled in layers on the bottom of the magma chamber, Skye, Scotland.

layers were deposited subaqueously on a sea or lake floor. However, in certain environments such as deltas, deserts or beaches, the bedding can be inclined relative to the depositional surface, for example cross-bedding in deltaic sandstones (see Figure 4.11b and accompanying text).

The form of the bedding surface is dependent on the environment of deposition and the compaction of the wet sediment during burial. Examples of the different forms of bedding surfaces include planar, wavy, lobate and irregular and are illustrated in Chapter 4. The change from one sedimentary rock to another may be gradational both vertically and horizontally, and therefore no well-defined discontinuity is present. The lateral extent of bedding surfaces is also very variable, some may be over 100 km in length, whereas others may be lenticular and very short. An assessment of the form and continuity of bedding surfaces to be expected within any project area must be made in relation to the sedimentary environment in which they were deposited.

In high-grade metamorphic rocks, which have undergone intense pressures and temperatures, there is a segregation between the light-coloured minerals such as quartz and feldspar

(Si/Al-rich) and the dark-coloured minerals such as biotite and hornblende (Fe/Mg-rich), and in places partial melting of the rock may occur. Such metamorphic segregation imparts a distinct layering to the rock, for instance in banded gneisses (Figure 9.4b). Although this layering may be planar over several kilometres, as seen in the gneissic banding displayed on the satellite image of Aswan, Egypt (Figure 9.2c), it is common for the layering to become highly disrupted, brecciated and folded during later tectonic and intrusive processes (Figure 9.3a). The characteristics of gneissic banding are discussed later in the chapter under the 'Gneissic foliation' section.

In relatively rare circumstances, igneous rocks will display compositional layering due to the varied concentrations of certain minerals which have crystallized from the cooling magma within a magma chamber or thick sill (Figure 9.4c). The layering results from the settling of the heavier minerals, such as the (Fe/Mg) silicate minerals olivine and pyroxene and the accumulation of these minerals on the floor of the magma chamber. Differentiation of the early crystallized minerals into layers by the circulating currents within the magma chamber may also impart compositional layering, but this is more irregular and may form along the margins of the magma chamber. Silica-rich, extrusive volcanic rocks may also display thin, disrupted compositional layering as a result of laminar flow (Figure 6.20b).

JOINTS

Joints are planar or curviplanar discontinuities in rock that define surfaces of potential or actual fracture but have no displacement across them (Figure 9.1). However, it is common for some joints to display evidence for very minor amounts of movement, possibly related to recent ground movements, but not sufficient to be classified as faults. Joints are extensional fractures that form during the formation of the rock by compaction of a sedimentary sequence or cooling of a magma, by tectonic forces related to uplift, faulting and folding and by unloading of the rock by erosion. These types of joints are referred to as primary, tectonic and sheeting joints, respectively. It is not always possible to determine the exact origin or timing of formation of particular joints, except where they are subparallel to the regional fault pattern or topographic surface and then a generic relationship is indicated. However, irrespective of the origin of joints, they are one of the main features to be seen in rock faces, tunnel walls and in drill core, and their orientation, spacing and surface characteristics are, therefore, of vital components of many geotechnical investigations. Systematic surveys of joint sets are conducted on outcrops and excavated rock faces with the aid of grids or traverse lines and, if appropriate, using remote radar instrumentation. However, from a statistical perspective, such surveys may not be completely robust, as some joint sets may be underrepresented within the 2D grid, vertical borehole or along the 1D traverse line.

Primary joints

Primary joints include those that were developed during the original formation of the rock, for instance during compaction of a sedimentary sequence or cooling of a lava flow, sill or plutonic body. However, in many cases, these joints are overprinted by a later set of joints that obscure the origin of the earlier joints.

Extension parallel to the bedding in sedimentary rocks, due to pressures exerted by the overlying strata, compaction and dewatering, provides an environment that primary,

(a)   (b)

Figure 9.5 Primary cooling joints. (a) Columnar cooling joints in a 30 m thick, Mesozoic, pyroclastic flow. The hexagonal columns range in width from 1 to 2 m, Ninepin Islands, Hong Kong. (b) Joints developed subparallel to the upper, horizontal boundary of a granite pluton, which has been intruded into a sequence of light grey volcanic rocks, Anderson Road Quarry, Hong Kong.

sub-vertical joints can develop at right angles to bedding. The early stages of uplift and unloading of a sedimentary basin can also create extensional stresses within the sedimentary layers, therefore providing the conditions for early joint formation prior to any major tectonic event.

Cooling joints in lava and pyroclastic flows provide the most impressive manifestation of primary joints. In Scotland and Northern Ireland, the perfectly formed, hexagonal columns, tens of metres high, were formed by the progressive cooling of thick basalt lava flows (Figure 6.19a). Columnar joints also form in those pyroclastic flows which are thick enough to act as single cooling units (Figures 9.5a and 6.19b). Many centres of crystallization were initiated in the flows close to the contact with the cold rocks over which they were extruded. As solidification of the flows continued, there was progressive crystallization, which resulted in the formation of planar, vertical extensional surfaces on a regular, commonly hexagonal, pattern.

Cooling joints also occur close to the contact of intrusive igneous bodies, such as plutons, with the surrounding rocks. Here, the joints developed subparallel to the intrusive contacts, particularly near the roofs of the intrusions where cooling of the magma was greatest (Figure 9.5b). Parallel joints may also form in the country rock, but the intensity of cooling jointing rapidly decreases away from the contacts. The identification of definite cooling joints in plutonic rocks, such as granite, is not unambiguous as sheeting joints can have similar orientations in certain topographic situations.

Tectonic joints

Tectonic joints formed during the regional deformation of the rocks as part of mountain building events. Different joint orientations are geometrically related to the principle stress axes (Figure 9.6a), which may be associated with the development of folds and faults. The pattern of joints across a typical fold is controlled by the maximum compressive stress, which is orientated perpendicular to the fold axial plane (Figure 9.6b). The joints form

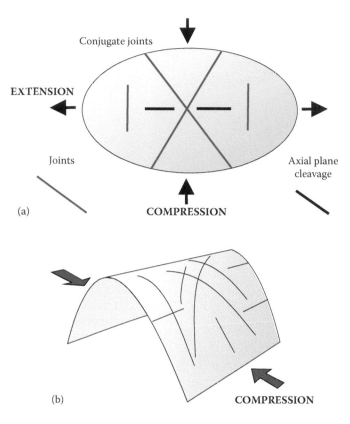

Figure 9.6 Tectonic joints. (a) Joint orientations related to the principal stress vectors and the strain ellipsoid. (b) Joints developed across a fold.

parallel and perpendicular to the strike of the beds, together with conjugate set of joints that develop in the planes of maximum shear. These orientations although initiated during a compressive regime are not developed as joints until the release of the stress in the crust during the waning of the tectonic event.

Sheeting joints

Sheeting joints develop in response to the unloading of the rock mass by progressive weathering and erosion of the near-surface materials (Figure 9.7a). They define low to moderately dipping slabs of rock that are subparallel to the present-day topographic surface (Figure 9.7b). Typically, sheeting joints are only encountered in the first few tens of metres of rock below the ground surface and commonly display a broad wave morphology in contrast with the more planar tectonic joints (Figure 9.7c). In homogeneous rocks such as granite, sheeting joints can take the form of domes, where successive layers of rock exfoliate around a hilltop (Figure 9.7d). Because it is common for sheeting joints to have lower dips than the ground surface of a hillside, they can form release surfaces for slope failure (see Figure 11.20). For example, the basal release surface of a fatal landslide in Hong Kong (Case Study 6.1) was partially controlled by a clay-filled sheeting joint in weathered volcanic rocks.

Figure 9.7 Sheeting joints. (a) Cross-section of a deeply eroded land surface illustrating the development of sheeting joints in relation to the topographic profile. (b) Series of low-dipping sheeting joints in granite on the flanks of Mount Kinabalu, Malaysia. (c) Wavy sheeting joints in granite near to the summit of Hwashan, China. (d) Exfoliation dome in granite, Bald Mountain, Texas, United States.

JOINT CHARACTERISTICS

Joints are of critical importance to many engineering projects as they greatly affect the engineering characteristics of the rock mass in quarrying, tunnelling, excavation and slope stability projects. They also largely determine the rates of water flow through a rock mass. It is, therefore, essential that joints are measured and described in detail with particular attention being placed on their orientation, spacing, persistence, surface features and infilling material as prescribed in standard texts (e.g. British Standards Institution, 1999).

Joint sets

Joints regularly occur in sets which have similar orientations and spacing (Figure 9.8a); however, where the joints do not belong to any well-defined set, they are referred to as *non-systematic joints* (Figure 9.8b). *Orthogonal joints* intersect at right angles (Figure 9.8c), whereas those with intersection angles of between 30° and 60° are called *conjugate joints* (Figure 9.8d).

(a)

(b)

(c)

(d)

Figure 9.8 Joint sets (a) Three, preferentially eroded, joint sets cutting low-dipping, sedimentary rock sequence, Ping Chau, Hong Kong. (b) Non-systematic joints in volcanic rock cut by pegmatite dyke. The joints post-date the intrusion of the pegmatite dyke which has been cut by small fault, Cape D'Aguilar, Hong Kong. (c) Orthogonal joints in a limestone bed, Dunraven Bay, Wales. (d) Conjugate joints in layered schist, Pos Selim, Malaysia.

It is not uncommon for several sets of joints to be of regional significance and to be encountered in many outcrops across an area. This consistency is related to regional uplift patterns or fault arrays that define the regional tectonic setting.

Joint spacing

The spacing of joints largely defines the size the blocks in a rock mass and therefore has significant control on the geotechnical parameters. The spacing and distribution of joints is controlled by many factors, including rock type, grain size, closeness to faults and position within the weathering profile. The spacing of joints is measured at right angles to the joint surfaces and is described extremely widely spaced (greater than 6 m) to extremely closely spaced (less than 20 mm). Figure 9.1 illustrates the contrasting joint spacing in granite between sub-vertical tectonic joints and more widely spaced, low-dipping, sheeting joints. In places, extremely close-spaced joints in an originally homogeneous rock, such as granite, can impart a laminated fabric to the rock (Figure 9.9).

Figure 9.9 Horizontal, extremely closely spaced, sheeting joints in vertical core in granite. Note uneven, rough and discontinuous form of the joints, Hong Kong.

The spacing of the joints may decrease towards a fault, and this may be more pronounced on one side of the fault than the other. This is well displayed in a granite quarry in Hong Kong where the joint spacing decreases markedly on one side of the fault (Figure 9.10a). The closer-spaced joints make the rock mass weaker and more susceptible to weathering over a much wider zone than the fault itself and, as a result, increase the negative topographic expression of the fault trace.

The joint spacing in a layered sequence may vary significantly, depending largely on the strength of the individual layers. For instance, in sedimentary sequences, the joints tend to be very well developed in the sandstone in comparison with the intervening mudstone beds where the intensity of joints is significantly less (Figure 9.10b). In folded sequences of stronger rocks, such as sandstone, the joints are not only more prominent but are also fanned around the fold hinge (Figure 9.10c).

The spacing of joints in drill core must be identified for each joint set. For example, a run of core through volcanic rock from Hong Kong (Figure 9.11) displays three joint sets with different joint spacing: a prominent moderately dipping set, a more widely spaced steeply dipping set and a weathered shallowly dipping set of sheeting joints. However, the recognition and determination of the spacing for joints orientated parallel or near parallel to the core axis can be misleading as they are not fully represented.

Joint surfaces

The surface features of joints and other discontinuities are described in terms of their smoothness, which can vary from smooth (Figure 9.12a) to rough (Figure 9.12b and c). These characteristics strongly influence the friction angle of the discontinuity. Description

(a)

(b)   (c)

Figure 9.10 Joint spacing. (a) Decreasing joint spacing close to a fault in granite, Mount Butler Quarry, Hong Kong. (b) Prominent closely spaced joints in sandstone layers in contrast with less well-developed and wider spaced joints in the tuffaceous siltstone layers, Tolo Channel, Hong Kong. (c) Joints in sandstone bed fanned around fold hinge, Anglesey, Wales.

of joint surfaces should follow prescribed nomenclature (e.g. Geotechnical Control Office, 1988; British Standards Institution, 1999).

Joint openings and infill material

The opening of joints is described from very tight to extremely wide, and this will largely determine the rate of flow of the groundwater though the joints. However, the size of the opening may vary naturally along the length of the joint. In samples from rock faces and

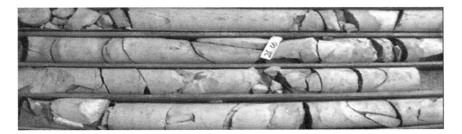

Figure 9.11 Three main joint sets in a vertical core run in volcanic rock. A prominent moderately dipping set, a more widely spaced steeply dipping set and a weathered shallowly dipping set of sheeting joints, Hong Kong. Length of core run 1 m.

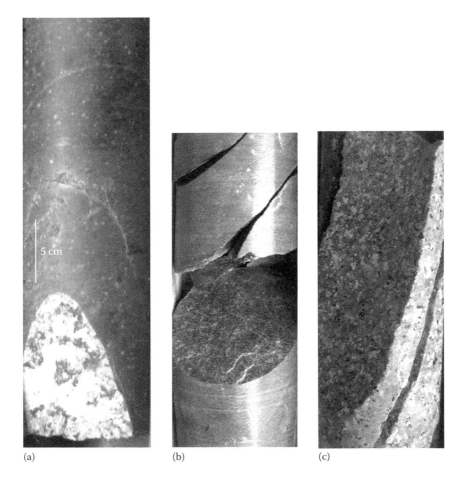

(a)   (b)   (c)

Figure 9.12 Examples of surface features of joints from vertical site investigation boreholes, Hong Kong. (a) Two extremely narrow, kaolin-coated, smooth joints in basalt. (b) Stepped joint surface in sandstone. (c) Closely spaced pair of very rough joints in granite.

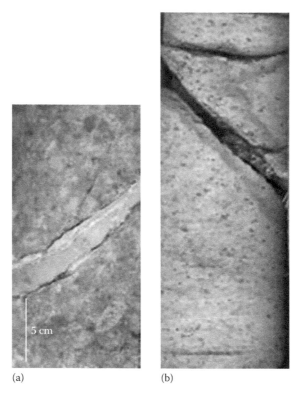

(a)   (b)

Figure 9.13 Infill material in joints from vertical site investigation boreholes in Hong Kong. (a) Calcite and chlorite infill cement in altered granite. (b) Sand infill in weathered joint developed in fine-grained granite close to ground surface.

drill core, the original aperture of the joints may be highly disturbed and therefore misleading. Joint openings may be filled by a variety of minerals prior to the weathering of the rock or as integral part of the weathering process. In the former case, minerals such as chlorite, calcite and quartz precipitated from hydrothermal fluids at depth (Figure 9.13a) and in the latter kaolin and manganese minerals precipitated from near-surface groundwater. In addition, very close to the ground surface, sand and clay can be washed into open joints (Figure 9.13b).

The composition of the infill material in joints is extremely important for certain engineering projects. For example, kaolin precipitation along sheeting joints greatly reduces their friction angle and the clay may act as an aquiclude if it forms a continuous layer. In both these cases, there is an increased risk of rock and/or soil failure if the orientation of the joints is adverse to the hill slope (see Case Study 6.1). In contrast, percolating hydrothermal fluids along the joints may precipitate minerals such as calcite and quartz that can effectively cement and strengthen the joint. In site investigations, it is important to recover the infill material in the joints, commonly, this material in rock is lost during the normal drilling process and specialized drilling techniques such as the use of polymer fluids may be required to recover the loose or weak material. However, joint infill material in soils is commonly well preserved in Mazier and standard penetration test (SPT) liner samples. It is also important to recognize that the rock adjacent to a joint may have been affected by groundwater flow or the passage of hydrothermal fluids along the joint. This may result in the selective alteration of the rock mass and as a result a change in its geotechnical properties.

FOLIATION

A *foliation* is any near-planar fabric in a rock and includes cleavage, schistosity and gneissosity and is restricted to fabrics that were imparted after the formation of the rock. In general, foliations are penetrative, very closely spaced, planar to wavy, tectonic discontinuities that form during episodes of folding or faulting. They occur in low- to high-grade dynamically metamorphosed rocks and can be found in deformed sedimentary, volcanic and igneous rocks.

Cleavage

Cleavage is defined by a set of extremely closely spaced set of discontinuities that generally develop in fine-grained sedimentary rocks during folding. Cleavage forms without the complete loss of cohesion of the rock and therefore differs from closely spaced fractures. Cleavage ranges from *slaty cleavage* found in low-grade metamorphic rocks (slates) thorough to *phyllitic cleavage* in medium-grade metamorphic rocks (phyllites). It is common for slaty cleavage to be orientated subparallel to axial planes of the folds (Figure 9.14a and b). The spacing of the cleavage is determined by the amount of strain the rock has undergone and the grain size of the original rock: fine-grained, clay-rich rocks develop planar, very closely spaced cleavage, whereas the cleavage in coarser-grained rocks is more irregular and wider spaced. Slaty cleavage (Figure 9.14c) is formed by the reorientation clay minerals and the formation of layer silicate minerals such as illite. Because of the preferred orientation of these minerals, the rock splits readily along planar domains into thin sheets which may be of sufficient quality to be used for roofing slates or most exceptionally as very smooth and planar slabs that form the base of some billiard tables.

Schistosity

As the metamorphic grade increases due to higher temperatures and pressures, a new set of coarse-grained micaceous minerals, such as chlorite, muscovite and biotite, crystallize.

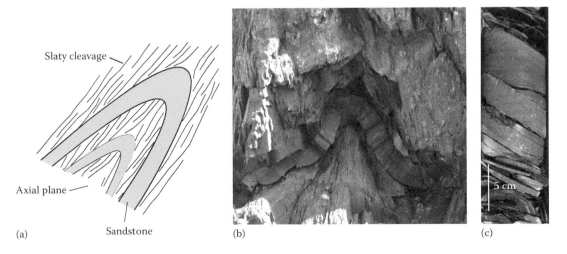

(a)   Sandstone   (b)   (c)

Figure 9.14 Cleavage. (a) Sketch of the development of cleavage subparallel to the axial plane of folds. (b) Cleavage in mudstone orientated subparallel to the axial planes of folds defined by sandstone layer, Anglesey, Wales. (c) Low-dipping slaty cleavage in mudstone, Vertical borehole, Hong Kong.

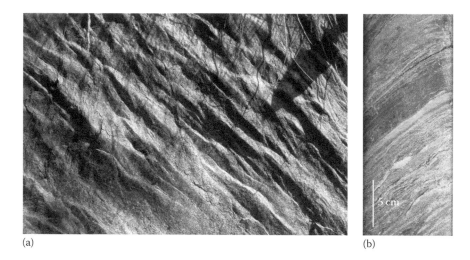

(a)   (b)

Figure 9.15 Schistosity. (a) Crenulated chlorite mica schist displaying shiny schistosity, Juneau, United States. (b) Wavy schistosity in metamorphosed calcareous sedimentary rocks. The darker bands are rich in biotite and define the original bedding in the rocks. The schistosity is subparallel to the bedding. Vertical borehole, Hong Kong.

These define a shiny, wavy foliation to the rock called *schistosity*, and the rock is referred to as schist (Figure 9.15a). In thin section, the schistosity may be seen to be defined by parallel-orientated flakes of biotite and muscovite (Figure 7.6a). The crystallization of new minerals is dependent of the composition of the original rock, and thus some layers may be preferentially rich in biotite (Figure 9.15b), whereas others may be devoid of this mineral. In some rocks, the schistosity is almost parallel to the bedding due to increased shearing in these medium- and high-grade metamorphic rocks.

Gneissic foliation

In the highest grades of metamorphism, the minerals are more equigranular and less orientated in comparison with schists, and thin layers of light and dark minerals define a *gneissic foliation* (Figure 9.16a). Commonly, pegmatite forms concordant or cross-cutting segregations and veins that developed by the partial melting of the rock at these high temperatures and pressures. Where the partial melt fraction becomes high, the rock is termed a *migmatite* (Figure 12.6a). Some gneisses are layered (Figure 9.4b) with bands rich in light-coloured minerals (quartz and feldspar), interspersed with bands rich in dark-coloured minerals (biotite, amphibole and pyroxene); these gneisses are referred to as *banded gneisses*. Where the banding is relatively planar and not folded, the gneisses may be split along the weaker, mica-rich layers (Figure 9.16b), but elsewhere the fracture potential of the rock takes no account of the orientation of the gneissic foliation (see Figure 9.3a). Homogeneous granites may develop into granite gneisses (Figure 9.16c) as a result of regional tectonism or local shear adjacent to faults; the primary minerals having been deformed to define a penetrative *gneissic foliation*. In places, the pegmatite fractions of the gneiss may be deformed into lenses that developed by shear and continued growth of the partial melt minerals. Where these lenses are large, the rock is called an *augen gneiss* (Figure 7.7d).

(a)

(b)   (c)

Figure 9.16 Gneissic foliation. (a) Detail of gneissic foliation defined by thin layers of light and dark minerals. Pegmatite segregations indicate partial melting of the rock, South Harris, Scotland. View width 10 cm. (b) Prehistoric standing stones formed of banded gneiss, Lewis, Scotland. The masons have used the gneissic foliation to split these 4 m high tablets. (c) Gneissic foliation in deformed granite from site investigation for bridge foundations, Vertical borehole, Hong Kong.

Mylonitic foliation

Close to and within ductile fault zones, the shear is so intense that a well-defined planar fabric is formed within and adjacent to the fault – this is referred to as a *mylonitic foliation* (Figure 9.17a). Mylonites are formed deep within the Earth's crust where temperatures and pressure are elevated and commonly partial melting of the rock has taken place. Within a mylonitic foliation, the quartz crystals become highly strained and elongated and, in places, the quartz crystals amalgamate into narrow ribbons. Other structures associated with mylonites, such as S–C fabrics are discussed further in Chapter 8. Mylonitic bands may be

(a)   (b)

Figure 9.17 Mylonitic foliation. (a) Narrow mylonite layer cutting deformed granite, Vertical borehole, Hong Kong. (b) Roofing tablets quarried from a thick mylonite zone formed close to the Main Boundary Thrust of the Himalaya. Note the faces of the tablets display a lineation that defines the direction of movement of the thrust, Nepal.

less than 1 cm wide (Figure 9.17a) but can be very significant in the ground model as they define distinct and laterally continuous planes of weakness. In areas where the shear deformation has been intense, the zone of mylonitic foliation may be several hundreds of metres wide, for instance close to one of the main thrusts related to the formation of the Himalaya during the collision of the Indian and Eurasian Plates. There, the dominant discontinuity in the rock is a very consistent, planar, mylonitic foliation, which has been exploited for the making of roofing tablets (Figure 9.17b). These all display a prominent lineation that developed parallel to the northward movement of the thrust.

CASE STUDY 9.1   SCHIST, DISCONTINUITIES AND FAULTS: LANDSLIDE, POS SELIM, MALAYSIA

[101°20.5′N 4°35.1′E]

Formation of a road in mountainous terrain near the Cameron Highlands hill resort in northern Malaysia resulted in extensive slope instability across a steep, forested hillside close to Gunung Pass (Figure CS9.1.1). In 2003, after extensive slope modification, the failure reached the ridge-line some 230 m above the road level, a 20 m high back scarp that developed together with a series of counterscarps within the upper part of the landslide, and a push-out zone at the base of the landslide just above the road (Malone et al., 2008). The 190 m high landslide has an estimated volume of approximately 2 million m³.

Figure CS9.1.1 Aerial photograph of the landslide. (Courtesy of A. Hansen.) The back scarp is located just above the drainage channel seen as a white line close to the top of the ridge.

GEOLOGY

The rocks of the Gunung Pass area consist of a sequence of Palaeozoic predominantly mudstones and sandstones that have undergone low- to medium-grade, dynamic metamorphism. These metasedimentary rocks outcrop in a 4 km wide, N- to NNE-trending, shear zone are flanked by Mesozoic granite intrusions (Figure CS9.1.2). Most of the central and upper parts of the hillside are composed of quartz mica schist with some quartzite layers up to 4 m thick

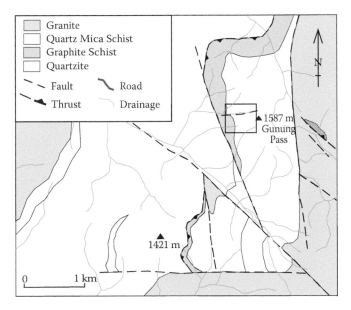

Figure CS9.1.2 Geology of the region in the vicinity of the landslide. The location of the map shown in Figure CS9.1.3 is indicated.

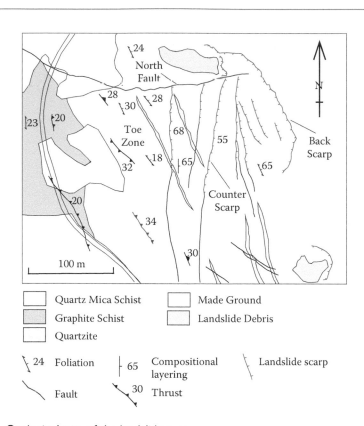

Figure CS9.1.3 Geological map of the landslide area.

and thin graphite schist layers (Figures CS9.1.3 and 9.1.4a). This schist unit is sharply underlain by a thick dark grey, graphite schist unit with thin interlayers of quartzite and quartz schist. All rocks display an intense penetrative foliation that is subparallel to the compositional layering, the bedding having been transposed during the deformation. Disrupted isoclinal fold hinges and S–C fabrics within the schists attest to the intensity of shear within this rock sequence. The foliation consistently strikes N–S and, in general, dips at shallow angles towards the east, into the slope. The schists are highly jointed with two of the sets orientated approximately perpendicular to the foliation (Figure 9.8d). Typical joint spacing is less than 0.5 m. Three fault arrays are present across the site. The first consists of low to moderately dipping thrust faults with similar orientations to the main foliation. These faults appear to have acted, in part, as release surfaces within the basal part of the landslide, and one is located with the toe zone of the landslide (Figure CS9.1.4b). The faults of the second array dip steeply towards E/ESE; they are slightly oblique to the main foliation traces and can be traced across the landslide. They have been reactivated within the landslide mass where they form counterscarps (Figure CS9.1.4d), which display both oblique and vertical striations, indicative of lateral and vertical displacements (Figure 8.12b). Only one fault belonging to the third array outcrops along the northern boundary of the site and defines the northern flank of the landslide (Figure CS9.1.4c). In the west, close to the road, the fault dips at 35°N, whereas to the east, higher

(a)   (b)

(c)   (d)

Figure CS9.1.4 Geological features of the landslide. (a) Compositional layering and foliation in the schist. (b) Thrust fault exposed close to the toe of the landslide. Very narrow fault is associated with the development of quartz veins. (c) The northern normal fault with reddened fault gouge. (d) Counterscarp formed in the upper part of the landslide during the downslope movement of the displaced block. Note slickensides on the re-reactivated fault plane. For close-up, see Figure 8.12b.

up the slope, it has increased to over 75°N. The bounding surfaces of this fault are very sharp (Figure 8.10a), and the low-dipping foliation in the adjacent schists has been steepened against the fault. The fault is over 2 m wide and composed of red-brown clay (gouge) with fine quartz gravel and buff silty clay with angular schist fragments (Figure 8.10a). The main surface features of the landslide are the main scarp, the head graben, the north and south flanks, the counterscarps of reactivated faults that cut obliquely across the site and a low-angle, push-out structure at the toe zone (Figure CS9.1.3). It has been suggested that at the head and in the upper main body of the landslide, the slip was on joints roughly orthogonal to the foliation, whereas in the toe area, the landslide mass slid upwards and out of the slope on along foliation planes associated with thrusting.

ENGINEERING CONSIDERATIONS

- The site is composed of schist within a major shear zone.
- Compositional layering and the foliation generally dip shallowly into the hillside.
- Several joint sets roughly orthogonal to the foliation and compositional layering.
- Three persistent faults cut obliquely across the slope and were reactivated during the failure.
- There has been no catastrophic failure, but there is bulging at the toe of the landslide along foliation and shear faults.

CASE STUDY 9.2   FOLDS, CLEAVAGE AND FAULTS: CUT SLOPE, GLANDYFI, WALES

[52°33.9′N 3°54.5′W]

I. Bews
Ramboll, U.K.

The widening and realignment of the A487 trunk road at Glandyfi, West Wales (Figure CS9.2.1), involved the excavation and stabilization of two rock cuts: one 450 m long with a maximum height of 15 m and the other 200 m long and 12 m high (Bews, 2012). The original slopes typically comprised patches of variably thick colluvium overlying weathered bedrock, overlying fresh, intact bedrock, with steep outcrops of bedrock. The colluvium is a granular–cohesive material, the source of which is weathered bedrock and the strength and stability of which derive from cohesion of the clay matrix and a degree of interlock from the included gravels and cobbles. The original profile of the rock face was largely dictated by the geology: zones where sandstone predominated and/or the bedding and cleavage geometry was favourable for slope stability tended to stand at steeper gradients (Figure CS9.2.2) than the more fissile, mudstone beds and those areas where the discontinuities are unfavourable to slope stability. The slopes were excavated in platforms, starting from the top and working down. The anchoring and concreting operations

Figure CS9.2.1 View from the top of one of the cut slopes over the River Dyfi and the A487 corridor.

Figure CS9.2.2 Steeply dipping sandstone and mudstone adverse to the cut slope with toppling of strata close to the original ground surface. Remedial works to stabilize the slope included rock anchors and reinforced, sprayed concrete retaining walls.

were completed before each platform was removed to continue on the next level down. Once the full-height concrete wall was complete, the masonry facing was constructed, tied in to the concrete at regular intervals.

GEOLOGY

The geology of the site and its surrounding areas (Figure CS9.2.3) was broadly known from the published 1:50,000 scale map (Sheet 163, British Geological Survey, 1989). The bedrock comprises sequences of thinly bedded Silurian sandstone, siltstone and mudstone (Figure CS9.2.4a), which have been folded into broad NNE-trending anticlines and synclines. Two significant mineralized faults pass through the site and are associated with zones of deformation including the development of kink bands (Figure CS9.2.5). The folds are associated with steeply to moderately dipping slaty cleavage (Figure CS9.2.4b and c) that is approximately parallel to the axial planes of the folds. The fold axial traces, identified on the geological map by fold closures, are oblique to the cut slopes (Figure CS9.2.3), thereby resulting in variable bedding and cleavage orientations, which in places adversely dip out of the slope. The exploitation of structural weaknesses in the bedrock by water and gravity has led to zones of relatively deep weathering producing clayey gravel with cobbles and occasional boulders. The rockhead profile was depressed by approximately 4 m above one of the faults. Here, the weathered material had been partly eroded away and the depression subsequently filled with colluvium. To ensure that the slope remained stable, this colluvium was locally excavated and supported by a gabion wall, which was subsequently covered by a masonry-faced retaining wall.

The potential for wedge and plane failure related to the orientations of the bedding and cleavage was of major concern during the design and excavation of the two rock cuttings for the new road alignment. Planar, wedge and toppling modes of failure were analyzed in detail for each

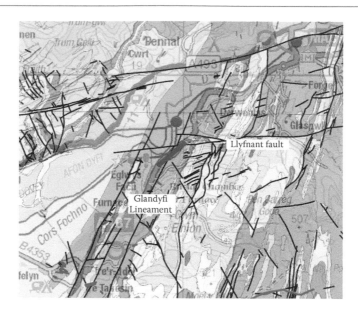

Figure CS9.2.3 Geological map of the site area and the site identified in red. (After British Geological Survey, Aberystwyth 1:50,000 Sheet 163, Solid and Drift Edition, 1989.) Note the orientations of the fold axial traces, which can be traced by connecting successive fold closures, and the fault arrays. The rock formations, defined by variations in mudstone compositions and sandstone content, are colour coded.

(a)   (b)

Figure CS9.2.4 Variations in the orientations of bedding and cleavage from different parts of the cut slopes. (a) Prominent bedding surfaces in sequence of mudstone/sandstone dipping out of the natural slope. (b) Low-dipping finely bedded mudstone cut by steeply dipping cleavage – both adverse to the slope.   (*Continued*)

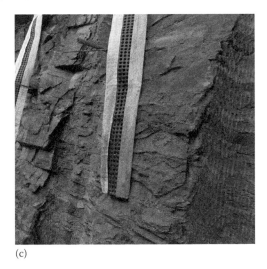

(c)

Figure CS9.2.4 (Continued) Variations in the orientations of bedding and cleavage from different parts of the cut slopes. (c) Adverse steeply dipping slaty cleavage cross-cutting finely bedded mudstone sequence that dips into the slope.

Figure CS9.2.5 Kink band in sub-vertical mudstone beds associated with faulting.

cutting. Global slope failure was also analyzed by effectively considering the rock slope as a soil mass, in which discontinuities in the bedrock may interconnect and form 'circular' slip surfaces. Figure CS9.2.4 presents three sections of the cut slopes with different orientations of bedding and cleavage. Where either or both of these planar discontinuities dip out of the slope, there was potential for failure.

ENGINEERING CONSIDERATIONS

- Geological map provides essential data for the geological model.
- Orientations of bedding and cleavage in relation to cut slope profile largely determine the potential for failure.
- Deep weathering above faults resulting in sharp depressions in rockhead.

Chapter 10

Folds

Folds are bends or buckles in layered or foliated rocks (Figure 10.1) that generally form in compressional tectonic environments related to the movement of crustal plates. They can affect a whole rock sequence across a wide area or be restricted to narrow zones, for example adjacent to faults or close to the intrusion of plutonic igneous bodies and salt domes. However, some folds are not tectonic in origin, but form within a wet sediments shortly after their deposition by slumping on a sloping seabed or lake floor (Figure CS9.2.4c), or can be found in recent glacial sediments related to the movement of ice. Folds can occur on various scales, and in general, the minor structures seen in hand specimen, core or even rock thin sections commonly mirror the larger-scale structures. Thus, the presence of small folds in boreholes will indicate the presence of more significant folds with similar origins, geometries and orientations.

The components and distribution of folded sedimentary rocks in tectonic zones can be studied in many parts of the world on satellite images using satellite images (Figure 10.2). These images can be vertically exaggerated, thereby enhancing the fold definition as the bedding appears steeper. In addition, using inclined views of the folds, it is possible to readily comprehend the relationships between the rock structures and the landforms. Figure 10.2a shows an area in the Peruvian Andes where the sedimentary layers have been folded into a south-east-trending open syncline in which the beds dip towards each other on opposite limbs of the fold. It is possible to trace some of the individual beds around the fold hinge. Comparable fold geometries are to be seen in a tectonic fold belt in Iran (Figure 10.2b), where a series of south-east-trending anticlines, in which the beds dip away from each other on opposite limbs of the fold, are separated by tight synclines.

From a geological perspective, folds can also provide evidence for the regional stress regimes acting during one or more tectonic episodes. For further information, the reader is referred to texts that describe the geometry, classification and kinematic analysis of folds in more detail (Fletcher, 1978; Ramsey and Huber, 1987; McClay, 1991; van der Pluijm and Marshak, 2004).

ENGINEERING CONSIDERATIONS

Folds are an essential part of understanding the geological evolution of an area as they provide evidence for the orientation of the regional stress fields during particular tectonic episodes. They commonly control the outcrop patterns at ground surface as observed on geological maps and satellite images and the subsurface distribution of the rock layers as constructed from surface measurements, site investigation data and geophysical surveys. An understanding of the form, geometry and orientation of folds is essential in developing the

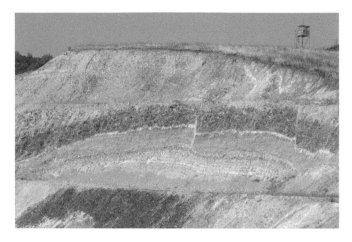

Figure 10.1 Open anticline in a sedimentary sequence intruded by a basalt sill exposed in a cut slope during formation of development platform. The fold hinge has been cut by a series of small normal faults, Western Turkey.

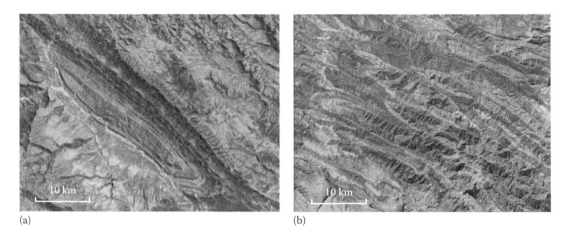

(a)   (b)

Figure 10.2 Large-scale fold structures seen on Google Earth, Landsat images. (a) Syncline in sedimentary rock sequence, Andes, Peru. The south-east-trending fold axis changes its plunge direction across the area to give the structure a basin-like form. The north-eastern fold limb dips steeply towards the south-west, and the south-western limb dips shallowly towards the north-east [18°1′S 65°44′W]. (b) Series of north-west plunging open anticlines with tight intervening synclines in a sedimentary rock sequence, Central Iran [33°0′N 48°37′E].

geological model for an area where folded rocks are present and predicting the distribution of the rocks beneath the ground surface and the attitude of the rock layers at the surface. Small folds, even at the core sample scale, can provide an indication of the form and structure of larger folds.

In order to formulate the most complete and robust geological model of site where folding has been recognized, it is imperative that the form and size of the folds and their origins are thoroughly understood. As fold structures may not be easily observed on the scale of a project site or in boreholes, it is essential that the regional context of the site is appraised in order to make reliable and geologically sound cross-sections, which can be used to explain the distribution of the rock layers, determine the orientation and nature of the rock layers and provide predictions for the proximity of faults. The distribution of distinct layers (either sedimentary or volcanic) with adverse geotechnical properties within a folded rock sequence can be predicted, if the fold geometry is defined from borehole intersections and rock outcrops within and adjacent to the site under investigation. Changing attitudes of rock layers due to folding can result in parts of a sequence becoming unstable on hillslopes, affect water flow patterns and result in different depths of weathering. The key engineering considerations that should be addressed for folds are as follows:

- Folds provide the geological framework of the project area and are an essential component of any cross-section.
- Across folds, the orientations of discontinuities, including bedding, compositional layering, joints and foliations, are variable but systematic.
- Rock layers are repeated over folds in a predictable manner.
- Zones of folding can indicate the proximity of faults, shear zones, igneous intrusions and salt domes.

Case Study 10.1 from a slate quarry in North Wales describes the geometry of a folded sequence of silty mudstone and siltstone that has been subjected to low-grade dynamic metamorphism. A well-defined cleavage has developed subparallel to the fold axial planes, and in certain stratigraphic units, a high-quality slate has developed. Quarrying of the slate has been undertaken for over a hundred years, and they are still being won from an extension to the old quarry further up the hillside (Figure CS10.1.1). The quarry is orientated along the axis of an open syncline, and the individual units may be traced around the quarry faces. A major concern of the quarrying operation is the risk of rock face failure, which is determined by the changing orientation of the discontinuities relative to the quarry faces. Reference should also be made to Case Study 6.1 where the geometry of folds and the location of the fold axial planes in a weathered volcanic sequence was an important factor in the development of a landslide.

FOLD NOMENCLATURE

Folds are made up of several distinct parts that provide the basis for their description and measurement in the field and on aerial photographs or satellite images. Figure 10.3 displays the main components of a typical fold. The *fold hinge* is the zone of maximum curvature on a fold and defined by the fold axis. It is flanked by *fold limbs*, which are essentially planar. The *fold axial surface* bisects the angle between the fold limbs and connects the fold hinges for successive layers in the fold. It is more commonly referred to as the *fold axial plane* as in most cases the surface is nearly planar. Although this is a geometric surface, it can be subparallel to a planar discontinuity or foliation formed during the fold episode (see Chapter 9). The fold limbs and fold axial planes are measured using dip and dip azimuth, whereas trend and plunge are used for the fold axis, which is a linear element. Statistical analysis of multiple readings of bedding across an area of folding can provide an accurate evaluation of the orientation of the fold axis. Such data can be an essential component of the construction of cross-sections and provide a reliable basis for the prediction of the geology in poorly exposed terrains.

The *core* of a fold is the central part of a fold and is flanked by the fold limbs. Where the limbs dip away from the fold axis and the core contains the oldest rocks, the fold is referred to as an *anticline* (Figures 10.2a, 10.3 and 10.7a). When the limbs dip towards the fold axis and the core contains the youngest rocks, the fold is said to be a *syncline*

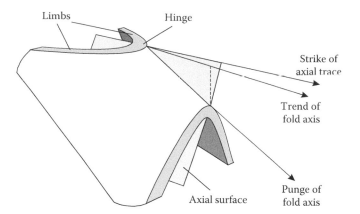

Limbs   Hinge

Strike of
axial trace

Trend of
fold axis

Axial surface

Punge of
fold axis

Figure 10.3 Geometric features of an anticline. *Fold limbs* – the sides of a fold. *Fold hinge* – point of maximum curvature on a folded surface. *Fold axis* – the line connecting the hinges of the folded surface. *Fold axial surface* – the surface that passes through the fold axes of successive folded layers. Where this surface is planar, it is referred to as the *axial plane* of the fold. *Fold axial trace* – the line where the axial surface intersects the ground surface.

(Figure 10.2b). The terms *synform* and *antiform* are used when the stratigraphic order in the fold is unknown, for example in high-grade metamorphic gneisses (Figure 10.6b), or when the beds have been transposed (see the 'Transposition structures' section).

FOLD CLASSIFICATION

Folds may be classified into three broad groups that are defined by the shape of the layers through the fold profile and the mechanism for their formation. The first group are called concentric folds which are generally found in gently deformed rocks and tend to be of only local significance, whereas the second group of folds, similar folds, are of more regional importance and commonly found in crustal-scale deformation belts. The third group of folds, called flow folds, are restricted to high-grade metamorphic terrains composed of gneisses in ancient cratonic complexes.

Concentric folds

Concentric folds form by bending of the strata and interstratal slip along the layer boundaries so that each folded layer is concentric to one another and the thickness of each layer, measured perpendicular to layer surface (t), remains constant (Figure 10.4a). This style of deformation can be demonstrated by bending a stack of printing paper and noting how the papers slip to form the fold. However, concentric folds across a layered sequence of rocks can only form as gentle warps, because as the fold becomes tighter, there is a space problem in the cores of the folds (Figure 10.4b). Here the thin, harder layers (e.g. sandstone) become highly crumpled, and the weaker material (e.g. mudstone) is squeezed out from the core of the fold (Figure 10.4c and d). As a consequence, concentric folds cannot propagate throughout a thick sequence of layered rocks; rather they are restricted to narrow zones or within particular layers. Commonly in concentric folding, the geometry and size of the folds in adjacent layers of different thicknesses are not identical (Figure 10.4e), and in these situations, the folds are classed as *disharmonic folds*.

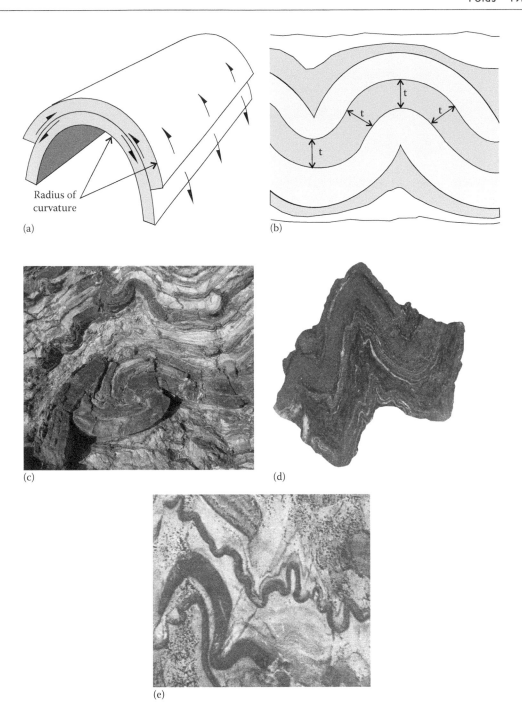

Figure 10.4 Concentric folding. (a) Formation of concentric folds by interstratal slip. (b) Cross-section of concentric fold showing the maintenance of strata thickness (t) around the fold hinge. (c) Concentric fold in sandstone and mudstone sequence. The sandstone beds keep their approximate thickness around the fold hinges, Zambujeira, Portugal. (d) Cut section through concentric fold – the sandstone bed is the same thickness across the fold, whereas the laminated shale in the core of the fold is highly distorted, Ogcheon, South Korea. Height of sample 12 cm. (e) Disharmonic folds in bedded limestone sequence. The bed thickness of the harder limestone layers (darker coloured) remains fairly constant across the outcrop, Rocky Mountains, Canada. View width 20 cm.

Similar folds

The second group of folds are *similar folds* in which the folding is accomplished by slipping along planar discontinuities (e.g. cleavage/foliation) parallel to the axial plane of the folds (Figure 10.5a). This style of deformation can be illustrated by pushing a finger into the end of pack of cards – slip occurs along the card surfaces and the ends of the pack form the shape of a fold (Figure 10.5b). In similar folds, the layer thickness measured perpendicular to the layers varies from a maximum in the core of the fold to a minimum on the limbs. However, the thickness of each layer measured parallel to the fold axial plane (T) remains constant. Similar folds commonly are associated with the development of closely spaced, planar discontinuities, which are orientated subparallel to the fold axial surfaces. Such discontinuities start to form during the early stages of dynamic metamorphism. Similar folds are characteristic of fold belts, and they can propagate throughout a complete rock sequence as there is no space problem in the cores of the folds (Figure 10.5c and d).

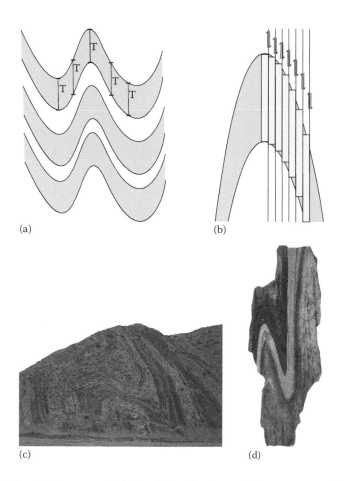

(a)

(b)

(c)

(d)

Figure 10.5 Similar folds. (a) Geometry of similar folds. The layer thickness measured parallel to the axial plane (T) is constant around the whole fold. (b) Schematic representation of the formation of similar folds by slip along planes parallel to the axial plane of the fold. (c) Similar folds in bedded sandstone/mudstone sequence. Note the folds are coherent throughout the folded sequence and that there is no disturbance of the layers in the fold core, Himalaya, Tibet. (d) Cut section through similar folds in bedded calcareous mudstone sequence. Note the bed thickness measured perpendicular to the bedding (t) is greatest around the hinges of the folds and least on the fold limbs, Ogcheon, South Korea. Length of sample 15 cm.

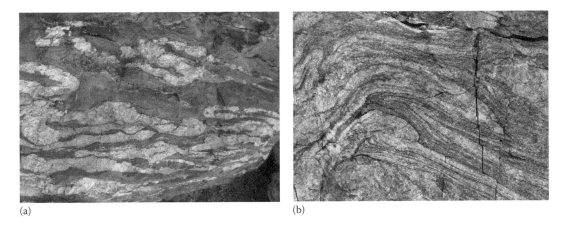

(a)   (b)

Figure 10.6 Flow folds. (a) Highly disrupted and folded granitic layers in a gneiss. The geometry of these flow folds is not constant across the outcrop, Himalaya, Tibet. (b) Refolded, isoclinal folds formed by progressive deformation at high temperatures and pressures, South Harris, Scotland.

Flow folds

Flow folds, as the name implies, develop where the rocks deform plastically, and the folds are very fluid in shape and have inconsistent geometries. The appearance of flow folds is akin to the folds seen in disturbed films of oil on the surface of water, where the folds swirl in all directions and their geometries are only consistent over a very small area. Flow folds are characteristic of very high-grade metamorphic zones where the temperatures and pressures at the time of the fold formation were close to the melting point of the rock. The flow folds within the high-grade metamorphic rocks in the shadow of Qomolangma (Mount Everest), Tibet (Figure 10.6a) display complex fold patterns, disrupted fold limbs and isolated fold hinges of pale-coloured granitic layers formed by the partial melting of the rock. In places, it is possible to distinguish folds of different generations that were formed as part of a continuum of the deformation process. For example, banded gneisses from the Isle of Harris, Scotland, contain isoclinal folds that have been refolded about axial planes almost at right angles to the first phase folds (Figure 10.6b).

FOLD GEOMETRY

Fold geometry is defined by fold shape, fold attitude and fold form. Together these provide information as to the tectonic development of an area and thereby will influence the formulation of the geological model, which in turn could have a bearing on the ground and design models.

Fold shape

The *fold shape* is defined both by the angle between the extension of the fold limbs, defined as the *interlimb angle* (Figure 10.7a), and the angularity of the hinge zone, which varies from rounded to angular. By definition *open folds* have interlimb angles between 10° and 70°, *close folds* have interlimb angles between 17° and 30°, *tight folds* have interlimb angles less than 30°, and the limbs of *isoclinal* folds are approximately parallel. The shapes of folds may vary within a single fold as illustrated in the open anticline in Anglesey,

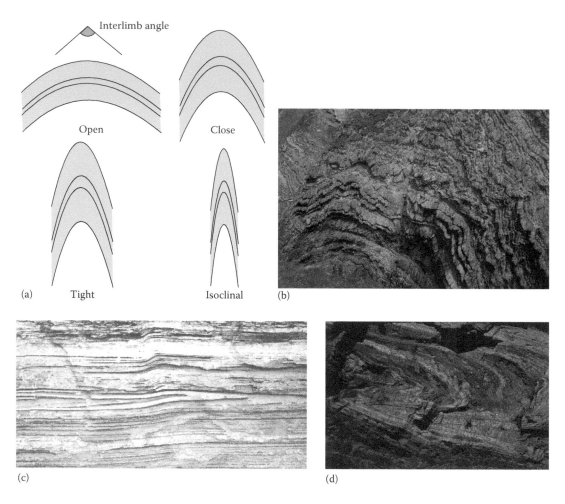

Figure 10.7 Fold shapes. (a) Sketches of different fold shapes. (b) Open fold in layered schist with subsidiary asymmetric close folds, Anglesey, Wales. (c) Isoclinal fold in layered limestone sequence which has been cut by a kink band, Ogcheon, South Korea. View width 60 cm. (d) Two tight folds with different interlimb angles and interlimb angles in mylonite associated with major thrust zone. The fold pair has a 'Z' shape, Hindu Kush, Pakistan. View width 2 m.

Wales (Figure 10.7b), which has numerous subsidiary close folds along its limbs. Isoclinal folds form in zones of maximum shortening of rock sequence during periods of the most intense deformation, such that the orientations of the fold limbs are almost the identical (Figure 10.7c). In this situation, rock layers may be repeated several times within a seemingly normal unfolded sequence. In Figure 10.7c, the layered sequence displayed on the left-hand side of the photograph is not in its original depositional order but contains layers repeated by the isoclinal fold. Where angular folds are confined to distinct bands, the folds are referred to as *kink bands* (Figures 10.7b and CS9.2.4). The interlimb angles of the fold hinges may also vary between adjacent fold hinges, for example, the folds in a mylonite from a thrust zone in the Himalaya (Figure 10.7d) have different shapes: a lower tight fold (interlimb angle of 20°) and the upper close fold (interlimb angle of 35°).

Observations on fold shapes across an area provide essential clues as to the intensity of deformation, degree of shear, proximity of major faults and extent of shortening during tectonic episodes, the knowledge of which will enhance the reliability of the geological model.

Fold attitude

Fold *attitude* is described in relation to the orientation of the fold axial plane and the plunge of the fold axis. The dip of the fold axial plane can vary from near vertical in upright folds, to moderate dips in inclined folds, to shallow dips in overturned folds, where one of the limbs is inverted, and lastly to near horizontal in recumbent folds (Figure 10.8a). Examples of folds with different attitudes are provided: upright folds (Figure 10.8b), inclined folds (Figure 10.8c) and recumbent folds (Figure 10.8d). Such variations reflect the intensity of the deformation and the orientation of the stresses that caused the folding. For example, in upright folds, the principal stress is roughly horizontal and reflects a compressional tectonic regime, whereas in recumbent folds, the principal stress is at a shallow angle to the vertical in response to shear.

The attitude of a fold is also described in relation to the plunge of the fold axes which are generally sub-horizontal to gently inclined (Figure 10.2), but where the folds have been

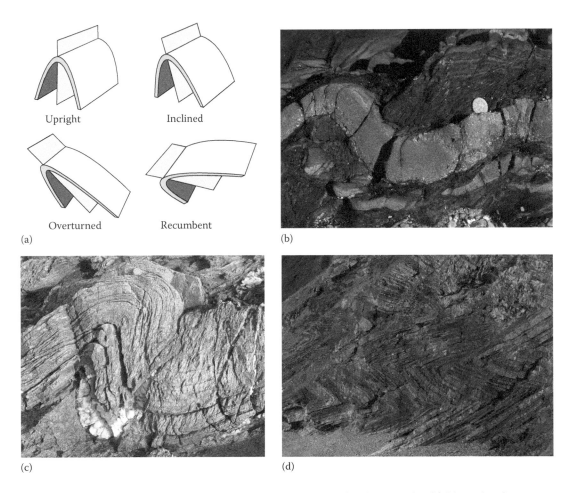

(a)   (b)   (c)   (d)

Figure 10.8 Fold attitudes. (a) Sketches of the descriptive terms for the attitude of folds with reference to the orientation of the fold axial plane. (b) Upright folds with vertical axial planes with rounded fold hinges in sandstone/mudstone sequence, Rhosneigr, Wales. (c) Inclined fold in sandstone/mudstone sequence where the axial plane is inclined and one of the fold limbs is vertical, Rhoscolyn, Wales. (d) Recumbent folds with horizontal axial planes and angular fold hinges, Devon, England.

reorientated during later phases of movements or in some fault zones with a strike-slip component, the fold axes may be steeply inclined or, in rare circumstances, vertical.

Fold form

Fold form is determined by the relative lengths of the fold limbs which if not the same give the folds an asymmetric shape (Figure 10.9a). Such fold asymmetry is dependent on the sense of shear, which varies across a fold from Z shape on one limb of a fold to M shape over the fold hinge to S shape on the opposing limb. The geometries of the small-scale structures can also provide data on the location of the larger fold axial planes. For example, the small folds seen in the open anticline from Wales (Figure 10.7b) show different asymmetries on opposite sides of the larger fold. On the left-hand side of the larger anticline, the minor folds have 'Z'-shaped symmetry, whereas on the right-hand side, they have 'S'-shaped symmetry – even without the larger anticline being exposed, one could locate the larger fold axial plane from isolated outcrop data. For example, the Z-shaped fold in a layered quartzite displayed in Figure 10.9b indicates that there is a recumbent antiform to the right of the

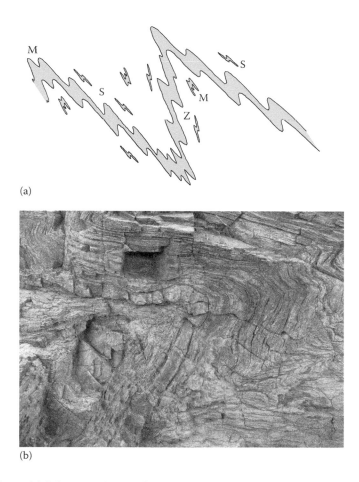

(a)

(b)

Figure 10.9 Fold form. (a) Schematic sketch of the different fold asymmetries. The minor folds have either Z or S shapes dependent on their position on opposite limbs of larger-scale folds. In the hinge zone, there is no asymmetry and the minor folds have M shapes. (b) Z-shaped fold in layered quartzites, Himalaya, Tibet.

photograph and a synform to the left. In rare instances, the form of small-scale folds can be recognized in site investigation rock cores, and recognition and description of these are important as they provide evidence for the style of deformation to be expected in the area (Figure 10.11a and b).

TRANSPOSITION STRUCTURES

In strongly deformed metamorphic rocks, the intense shearing of the original rock strata results in isoclinal folding, the attenuation of the fold limbs, the separation of the fold hinges into isolated hooks and the development of a strong foliation parallel to the compositional layering (Figure 10.10a). The layering in a transposed zone should not be referred to as bedding as the same layers may be repeated several times throughout a section with some beds being overturned and others the right way up.

(a)

(b)

(c)

Figure 10.10 Transposition structures. (a) Schematic diagram of the development of transposed bedding (S_0, bedding; S_1, secondary foliation). (b) Transposed sandstone beds in schist sequence exposed within a major landslide. The secondary foliation is subparallel to compositional layering, Po Selim, Malaysia. (c) Detail of the schist from the Malaysian landslide containing lenses of hard, sheared material with relict, isoclinal fold hinges.

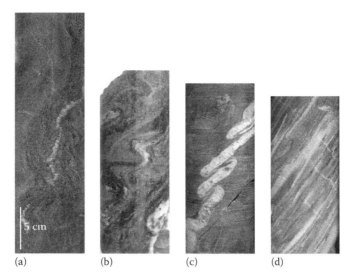

Figure 10.11 Minor fold structures in vertical borehole rock core, Hong Kong. (a) Open recumbent folds indicative of a thrust fault terrain. (b) Asymmetric angular folds in deformed schist. The asymmetry provides evidence for the predicted form and geometry of the larger-scale folds (see Figure 10.9). (c) S-shaped folds in a quartz vein in mudstone. One set of fold limbs have almost been totally sheared out. (d) Transposed compositional layering, highlighted by thin, pale grey marble layers, in a calcareous schist. The transposition has been intense and a few relict fold structures remain.

Identification of transposition structures through the presence of sheared out fold limbs within a very closely spaced secondary foliation subparallel to compositional layering (Figure 10.10b) and detached isoclinal fold hinges (Figure 10.10c) provide vital evidence of the deformation characteristics for an area, in particular the presence of shear zones, which may contain significant discontinuities with specific geotechnical properties, such as low friction angles. Transposition structures can also be identified in core samples; for example, one of the limbs of a folded quartz vein has almost been completely sheared out (Figure 10.11c), and the thin layering of a deformed calcareous schist (Figure 10.11d) has no stratigraphic significance as the isoclinal hooks and strong foliation suggest that many of the layers have been repeated during the transposition process. The recognition of these isolated fold hinge zones in the core is a vital part in the development of the geological model and cross-sections for a project area and will alert the geologist to the presence of shear zones and the geological complexity of site (see Case Study 9.1).

CASE STUDY 10.1   DISCONTINUITIES IN SLATE: SLOPE STABILITY IN QUARRY, BETHESDA, WALES

[53°9.7'N 4°4.0'W]

D.E. Jameson

GWP Consultants LLP, Chipping Norton, U.K.

Penrhyn Quarry is situated 1 km south-west of Bethesda, North Wales, adjacent to the Snowdonia National Park (Figure CS10.1.1). The working of slate has been in continuous operation for over 200 years. It remains the largest slate quarry in the world from which the famous Penrhyn Blue Slate has its origin. The historic workings were concentrated in the former North Quarry area (now flooded) and established the hillside quarry bench working method used in the majority of quarries today. The North Quarry area was abandoned in 1989 following a substantial sidewall collapse (estimated at 3.2 Mt) of the south-eastern face. This was then the latest of a series of historic failures along these faces. As a result of the fall, the workings in the northern half of the North Quarry were abandoned to concentrate on the development of the active South Quarry area. During the quarrying operation considerable quantities of waste material is produced. This consists of superficial deposits, weathered rock down to a depth of 65m, and the by-products of the slate production. Large waste tips flank the site and a new internal tip now buttresses the unstable south-eastern quarry faces.

GEOLOGY AND SLOPE STABILITY

The quarry is developed within the Llanberis Slate Formation of Cambrian age comprising a thick sequence of silty mudstone and interbedded siltstone with a stratigraphic thickness of over 600 m. The strata have been subject to low-grade regional metamorphism which has

Figure CS10.1.1 Inclined aerial photograph of site from the north.

Figure CS10.1.2 Geological features within the slate quarry. (a) Folded bedding cut by slaty cleavage. (b) Planar failure on shallowly dipping joint. (c) Active toppling instability along cleavage planes.

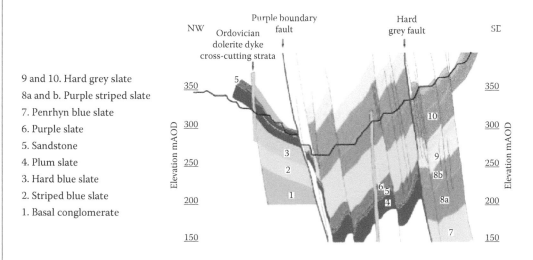

Figure CS10.1.3 Schematic cross-section of South Quarry.

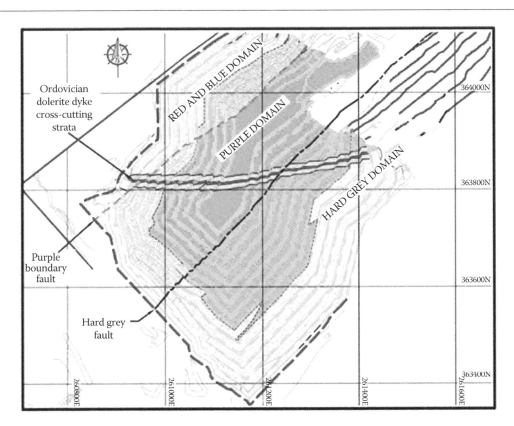

Figure CS10.1.4 Final South Quarry design, boundary fault positions and geotechnical domains.

induced a well-developed slaty cleavage dipping at 80°SE (Figure CS10.1.2a). The slates exhibit distinct but subtle colour variations and are occasionally interbedded by 'marker' sandstone units. The variation in slate colour is thought to arise from minor variations in the amount of iron in the clay minerals that constitute a large proportion of the original sediments. The structure within the quarry is dominated by a NE-trending, close syncline (Figure CS10.1.3) in which the cleavage is subparallel to the fold axial plane. The folded strata are transected by a series of faults, the majority of which dip steeply towards the SE.

The rock mass is cut by a number of discontinuity sets (e.g. Figure CS10.1.2b) which ultimately control the overall slope angle allowing for continued bench access. These include bedding that is generally seen as a faint lineation on cleavage surfaces, but in places as persistent thin soft clay or sandstone bands, cleavage, and at least six joint sets and faults.

The South Quarry is arranged over 12 benches resulting in an overall working quarry depth of 190 m. Geological and geotechnical domains comprise areas of similar geological materials and/or geological structure (Figure CS10.1.4). There are three principal domains separated from one another by the large 'boundary faults' which trend subparallel to the direction of quarry face advance. Discontinuity orientation and persistence combined with slope alignment and face height dictate the style and magnitude of potential instability. A kinematic assessment for potential planar, wedge and toppling failure using stereographic techniques on the various joint sets present in each geotechnical domain is used to design slopes to

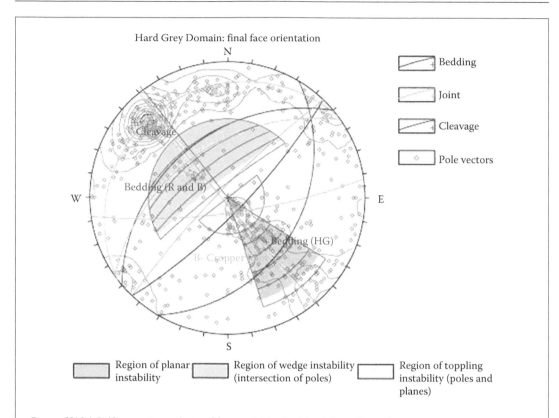

Figure CS10.1.5 Kinematic analysis of faces within the Hard Grey Domain.

eliminate or accommodate the potential for overall slope failure and ensure long-term access (Figure CS10.1.5).

The potential failure slope mechanisms vary between domains: (1) Hard Grey Domain (slope stability is dominated by toppling (Figure CS10.1.2c) influenced by the boundary fault position and cleavage orientation), (2) Purple Domain (minor wedge failures), and (3) Red and Blue Domain (the slope stability is controlled by the potential of planar instability on bedding).

ENGINEERING CONSIDERATIONS

- Complicated geological structure requires regular update mapping to determine the position of weak bedding parallel units.
- Geotechnical domains are defined by boundary faults across which the disposition of discontinuities and the style of the slope instability changes.
- Exposed face length is reduced by in-pit buttressing to increase long-term stability.

Chapter 11

Weathered rocks

Weathering is the process of alteration and breakdown of rock at or close to the Earth's surface by chemical decomposition, biological degradation and physical disintegration that leads to the formation of soil. Weathering is most severe in humid tropical climates (Figure 11.1a and b) where it may extend to over 200 m beneath ground surface, whereas in more temperate climates, the process is much slower and less intense and may only affect the upper few metres of rock. The depth of weathering is controlled by the length of time that the decomposition has been active. During weathering, the constituent minerals in the rock become altered to clays, iron oxides and a plethora of other secondary minerals. Physical disintegration of the rocks through annual wetting and drying cycles, temperature changes and surface abrasion and fracturing are additional factors that can influence the intensity, distribution and rate of weathering. Weathering also involves the dissolution of soluble minerals within some rocks and the precipitation of new minerals from percolating groundwater. The formation of cave systems in carbonate-rich rocks is part of the overall weathering process and leads to the development of karst.

ENGINEERING CONSIDERATIONS

As a consequence of weathering, rocks generally become weaker, more porous and in places more fractured than the original rock (Figure 11.2); however, in rare circumstances, weathered rocks may be exceptionally hard, for example silcrete which forms in tropically weathered soil profiles as a result of the precipitation of microcrystalline quartz. The understanding of the distribution and geotechnical characteristics of weathered rocks and soils is crucial for the engineering design of many projects, including building and dam foundations, cut slopes and tunnels (Saunders and Fookes, 1970, Martin and Hencher, 1986).

Within any particular climatic zone, the intensity and style of weathering, and therefore the geotechnical characteristics, will largely be dependent on the mineralogy of the rock and the type and spacing of discontinuities. For example, quartz-rich rocks (e.g. quartz sandstones, welded ash-flow tuffs) and massive rocks with few discontinuities weather more slowly and to a shallower depths than those rocks rich in feldspar or carbonate minerals (e.g. granite, limestone) and intensely jointed or foliated rocks.

In general, the intensity of weathering decreases with depth; however, the transition from rock to soil may be abrupt or gradational dependent largely on the hydrology of the site. In addition, it is not uncommon for bodies of rock to be surrounded by completely weathered material (see Figure 11.8), and in rare cases, thick slabs of rock may overlie intensely

(a)

(b)

Figure 11.1 Tropical weathering. (a) Deeply weathered sedimentary rock sequence exposed in a cut slope of the opencast Penjom Gold Mine, Malaysia. The intensity of weathering increases towards the ground surface, with an iron-rich zone at depth that passes upwards into a kaolinitic clay zone. Close to the ground surface, the original rock structures are completely destroyed within a residual soil horizon. (b) Deep gullies in deeply weathered granite. The white colouration is due to the high content of clay derived from the original feldspar and mica in the granite, New Territories, Hong Kong.

Figure 11.2 Weathered volcanic tuff at the portal to a drainage tunnel, Tsuen Wan, Hong Kong. The weathering, which is largely controlled by joints, is highlighted by iron staining and increases towards the upper part of the excavation. Close to the floor of the tunnel light grey relatively fresh tuff is exposed.

weathered material (see Case Study 11.1). All these situations need to be understood and addressed in the planning of the site investigation boreholes and fully taken onto account in the engineering design of foundations or other works.

The hydrogeology of weathered rocks can be highly significant in certain weathered profiles where the water flow can be channelled along pipes within the soil or above the soil/rock interface. These allow rapid transfer of rainwater from ground surface to deep within the soil mass, thereby reducing the stability of hillsides.

The key engineering factors which should be considered in any project where weathered rocks are present are as follows:

- In general, weathered rocks are weaker and more friable than fresh rocks.
- The distribution of soil and rock is very variable and controlled by geological features.
- The depth of weathering can range from less than a centimetre to over a hundred metres.
- The development of kaolin in the soil profile can reduce the shear strength of the soil mass, and where concentrated into seams may influence the hydrology of the site.
- Carbonate-rich rocks dissolve during the weathering process to form a complex terrain termed karst (see Chapter 12).

Two case studies are presented in this chapter, which illustrate the importance of understanding weathering processes and their implications for engineering projects. The first describes the initial site investigation for the foundations of school buildings in Hong Kong (Case Study 11.1). Deep weathering in granite associated with a moderately dipping shear zone was found to underlie part of the site, and as a result, footprints of the buildings were moved. The second case study involved the geological and geotechnical assessment of a railway alignment in Mongolia (Case Study 11.2), where weathering in the harsh cold environment had resulted in the rock substrate being deformed by frost action, cryoturbation and evaporite deposition in the softer residual soils. In addition, recent cementation of the superficial deposits has resulted in the formation of an iron-rich duricrust. Reference should also be made to the weathering of folded volcanic rocks presented in Case Study 6.1.

SAPROLITE

Saprolite is a term that is used for weak, friable, chemical-weathered material in which the original structure and fabric of the rock is preserved (Geological Society, 1997). For example, the textures, mineral shapes and discontinuities of the original rock are minutely preserved in the weak saprolitic soils: the crystal shapes and texture of granite (Figure 11.3a), the thin bedding in sedimentary rocks (Figure 11.3b) and the morphology of the ejected crystals in a volcanic tuff (Figure 11.3c). Saprolite is porous and susceptible to slow leaching of the minerals, so that silica, iron and aluminium pass into solution. The development of saprolite at any particular locality is dependent on several factors: the climate during its formation (e.g. humidity, temperature), the duration of the saprolite formation, the composition of the parent rock and the topography of the site which controls the transportation of the leachate (high in silica, bases and iron) in the groundwater and the erosion of the weathered material. Some saprolites are very old, possibly Tertiary in age, and have formed over many millions of years and may not relate to present climate conditions. Saprolite is generally well developed over quartz- and feldspar-rich rocks, whereas over ultramafic rocks with no quartz, the transition to soil is more abrupt.

DURICRUST

Duricrust is the residual accumulation of iron and alumina (laterite and bauxite) or the precipitation of minerals such as calcite (calcrete) and silica (silcrete) from groundwater. It is dominantly formed in tropical climates particularly over ancient landscapes where the duricrust may attain thicknesses of tens of metres; however, thin deposits of iron can occur

(a)

(b)

(c)

Figure 11.3   Examples of saprolite showing the original textures of the rocks preserved in the soil. In all cases, the saprolites are extremely soft and can be grooved by fingernail. (a) Medium-grained granite with the feldspar crystals completely replaced by clay minerals, Mount Butler Quarry, Hong Kong. (b) Mazier sample of completely decomposed calcareous schist, with intricate detail of the deformed foliation, Tung Chung, Hong Kong. (c) Completely decomposed crystal lithic tuff with iron-stained irregular fractures. The shapes of the quartz and feldspar crystals are preserved, Singapore. View width 10 cm.

as 'hard ground' in temperate soil profiles close to the water table. *Laterite* is the most widespread form of duricrust and consists of a red-brown, iron-rich deposit that is generally crumbly and porous (Figure 11.4b). It has been used as a building material for over a thousand years as it is relatively soft when wet and can be cut easily into blocks which harden when dry.

Figure 11.4   Laterite. (a) Cerro Manomo, eastern Bolivia, is composed of a dissected remnant of ancient laterite that developed on the Paucerna planation surface. It overlooks a regional undulating San Ignacio planation surface that is covered by a younger laterite. (b) Laterite building block from Suthokhai, Thailand. View width 5 cm. (c) Schematic section across eastern Bolivian Precambrian Shield showing the development of three planation surfaces of Tertiary age and their laterite (brown) deposits. Vertical scale exaggerated. (After Litherland, M. et al., The geology and mineral resources of the Bolivian Precambrian shield, Overseas Memoir of the British Geological Survey, No. 9, 1986, 152pp.)

In tropical mid-continental stable areas such as central parts of South America, Australia and Africa, the long weathering cycle has resulted to lateritic weathering profiles over 30 m in depth and in places over 100 m. These commonly form plateaux extending over thousands of square kilometres, and because of subsequent uplift, erosion and climate change, the older plateaux are left as isolated flat-topped hills capped by thick lateritic soil profiles. For example, Cerro Manomo in eastern Bolivia is a remnant of a once extensive laterite surface that would have covered a large portion of the Precambrian Shield area (Figure 11.4b). Across this region, four Tertiary laterite planation surfaces (Figure 11.4c) have been recognized (Litherland et al., 1986).

WEATHERING PROFILES

In tropical areas, the weathering profiles from fresh rock to completely decomposed material range from gradational to sharp. In relatively homogeneous rocks, such as granite, the most common gradational profile displays an upward passage from minor discolouration of the fresh rock along joints, through selective decomposition of the rock adjacent to the joints, then into completely decomposed material in which textures and structures of the fresh rock can be recognized and finally into residual soil close to the ground surface, which is devoid of any original textures of the rock (Ruxton and Berry, 1957). In more temperate climates, the weathering profiles are much thinner and staining and fracturing of the rock are the dominant characteristics, although at the ground surface, organic influences allow residual soils to develop.

Figure 11.5a displays a schematic representation of a typical tropical weathering profile in granite whose variations will be largely dependent on the orientation of the joints, particularly the low-dipping sheeting joints, the presence of faults and shear zones, distribution of hydrothermal alteration, development of impermeable clay-rich layers and the hydrogeology of the site. A tropical weathering profile in granite exposed in a road cut from Malaysia (Figure 11.5b) illustrates the gradation from fresh rock to soil over a vertical distance of about 20 m. In less homogeneous rocks such as layered sedimentary and volcanic rocks and foliated metamorphic rocks, the weathering will be additionally influenced by different compositions of the individual layers and the spacing of metamorphic fabrics. For example, the temperate weathering of a steeply dipping sequence of slightly metamorphosed interbedded sandstone and mudstone from Portugal (Figure 11.5c) displays different degrees and depths of weathering between the two rock types. Here the weathering zone is only a few metres deep, and decomposition of the sandstone is controlled by both the bedding planes and the joints. Close to the ground surface, the weathering is largely dominated by fragmentation of the rocks, rather than chemical replacement of the component minerals. The schematic representation of this situation is shown in Figure 11.5d where the jointed sandstone is stronger and less weathered than the fissile mudstone that has been weathered to greater depths.

In some situations, the gradation from rock to soil may be abrupt due to the hydrogeological conditions of the site. For example, the progressive development of an impermeable clay-rich layer along a low to moderately dipping sheeting joint or other discontinuity can impede the circulation of groundwater so that the weathering is largely restricted to the zone above this layer. Figure 11.6 displays a sharp rock/soil boundary in volcanic rocks from Hong Kong. Close to this boundary, within the soil, white clay-rich lenses and stringers can be seen. The formation of kaolin in tropical weathering profiles is covered in more detail in the 'Clays in the weathering profile' section.

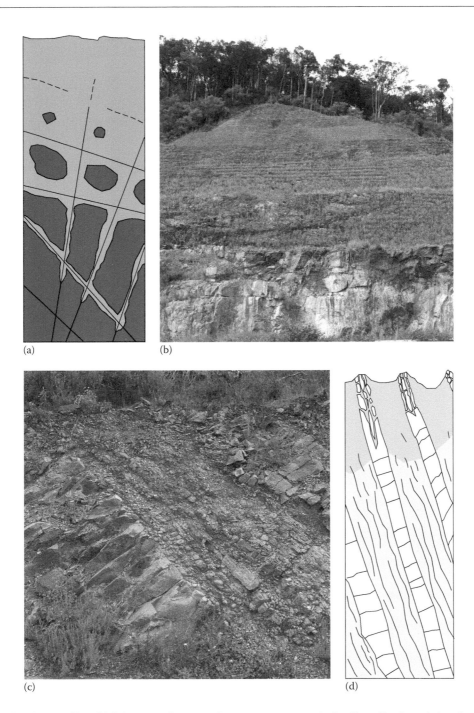

(a)   (b)

(c)   (d)

Figure 11.5 Soil profiles. (a) Schematic diagram of most common tropical soil profile above jointed granites (red – rock, pink – soil, brown – residual soil). (b) Gradational tropical weathering profile in granite from fresh granite (Grade I) to residual soil (Grade VI). Road cut in Cameron Highlands, Malaysia. (c) Thin temperate soil profile in bedded sandstone and foliated mudstone. The foliated mudstones are weathered to a greater depth than the more homogeneous and stronger sandstones. Road cut close to Santa Catarina, Portugal. (d) Schematic diagram of temperate soil profile in a sedimentary, moderately dipping, rock sequence (grey – mudstone, yellow – sandstone, brown – weathered rock).

Figure 11.6 Sharp contact between fresh volcanic rock and completely weathered, *in situ*, material. Note the presence of white clay layers close to the soil/rock interface, Kong Island, Hong Kong. (From Geotechnical Control Office, 1988. Published with the permission of the Civil Engineering Department, Hong Kong SARG)

Weathering grades

The progressive weathering of rocks results in a change of the geotechnical properties of the materials from strong fresh rock to weak soil. This change has been catalogued for different rock types under different weathering conditions, and the most appropriate scheme should be determined at the start of any project, so that geotechnical parameters may be assigned to the different grades identified. Here, we present the weathering grade classification for granite and volcanic rocks as used in Hong Kong and set out in Geoguide Number 3 (Geotechnical Control Office, 1988). This scheme is used to describe materials from site investigation boreholes and characterizes the geotechnical properties of the rocks and soils within the weathering profile. It divides the rock and soil into six grades, Grades I–VI which may be readily recognized in outcrop and in borehole and Mazier samples without the use of specialist equipment (Fookes and Horswill, 1970). The grades are identified by several distinguishing features: the strength of the material as defined by the ease of breaking the sample by hammer, hand or crumbling by finger pressure, the discolouration of the material (particularly iron staining) in comparison to the colour of the fresh rock and whether the sample slakes in water. Table 11.1 sets out the terms used in the description of weathered granite and volcanic rocks together with the grade symbol and distinguishing features. The most important boundary is that between Grade III and Grade IV for it marks the boundary between rock and soil. Boundaries between the grades are commonly gradational, but in places they may be sharp and distinct, and different grades may also be interleaved. Focus here is made on samples recovered during site investigations, for it is the core and Mazier samples that are most commonly seen by the engineer.

Figure 11.7 presents a sequence of core and Mazier samples recovered from site investigations in Hong Kong. They display the critical characteristics of each grade as described in the figure caption. The only criteria that cannot be determined visually relates to the boundary between Grades V and VI which relies on the latter slaking in water.

Table 11.1 Weathering grades of granite and volcanic rocks and their distinguishing features

Descriptive term	Grade symbol	Distinguishing features in granitic and volcanic rocks
Residual soil	VI	No original rock texture preserved Crumbles by finger pressure
Completely decomposed	V	Original rock texture preserved Crumbles by finger pressure Slakes in water Completely discoloured relative to fresh rock
Highly decomposed	I	Original rock texture preserved Brakes into smaller pieces by hand Does not slake in water Completely discoloured relative to fresh rock
Moderately decomposed	III	Rock completely stained Cannot be broken onto smaller pieces by hand Easily broken by geological hammer
Slightly decomposed	II	Stained close to joints Not easily broken by geological hammer
Fresh	I	No staining Not easily broken by geological hammer

Source: After Geotechnical Control Office, *Guide to Rock and Soil Descriptions* (*Geoguide 3*), GEO, Hong Kong
Government, 1988, 76pp.

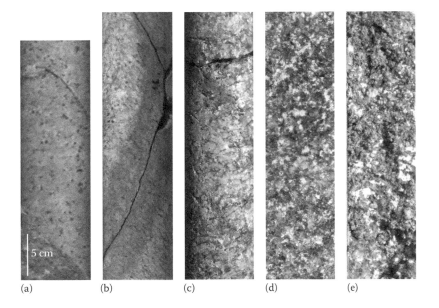

(a)   (b)   (c)   (d)   (e)

Figure 11.7 Weathering grades in granitic rocks from site investigation boreholes. Hong Kong. Samples
(a)–(c) are from rock core and samples (d)–(e) from Mazier samples. (a) Grade I, fresh granite with no staining. (b) Grade II, granite with iron staining of the rock adjacent to joints.
(c) Grade III, granite with pervasive iron staining and partially decomposed feldspar crystals.
(d) Grade IV, granite soil with complete decomposition of primary minerals. (e) Grade V,
granite soil with primary texture preserved and slakes in water.

Corestones

The initial decomposition of a homogeneous rock is mainly focussed along joints (Figure 11.8a) and continues by the progressive alteration of the rock away from the joint surface to produce a series of concentric fractures that are commonly referred to as 'onion skin' weathering patterns (Figure 11.8b). As weathering progresses, as seen in the upper parts of the weathering profile, the rock component is reduced although the relict joints may be identified the soil (Figure 11.8c). Finally, only rounded rock masses, called *corestones*, remain within a background of completely decomposed rock (Figure 11.8d). Corestones are a common feature of tropically weathered granites but may also form in volcanic, sedimentary and metamorphic rocks where discontinuities such as compositional layering or foliations may control the formation of corestones in addition to the joints.

The development of corestones is an important feature of the weathering profile as it will determine the distribution and size of rock masses within weaker decomposed rock (soil), thereby influencing the mass geotechnical characteristics of the ground. In boreholes, it is often difficult to identity corestones and determine whether rockhead (see next section) has been reached, as some corestones can be many metres across.

(a)   (b)

(c)   (d)

Figure 11.8 Progressive development of corestones. (a) Decomposition of a volcanic rock controlled by an orthogonal set of joints. Note development of white clay seams along relict joints. Nine Pin Islands, Hong Kong. (b) 'Onion skin' weathering patterns in jointed limestone, Phuket, Thailand. (c) Rounded corestones of syenite with relict joints visible in the completely decomposed material, Monchique, Portugal. (d) Isolated rounded corestone in completely decomposed syenite, Foya, Portugal.

Rockhead

Rockhead is an idealized surface where soil (Grades IV–VI) passes downwards into rock (Grades I–III). In places this surface can be extremely sharp, but more commonly, there is no discrete boundary between soil and rock; either there is a gradational transition between the two or there is a downward increase of rock masses within the soil profile. In the latter case, the rock masses may either be corestones or slabs of rock sandwiched between layers of soil. In borehole samples, it is impossible to distinguish between these two types of rock mass as their 3D form needs to be known.

The identification of rockhead is very important in the foundation designs that specify to be founded in rock. The misidentification of the position of rockhead may lead to substantial soil layers being present beneath the end of the pile or unnecessary extension of the pile depth. In Hong Kong, the normal practice is to define rockhead for foundations in granite and volcanic rock as the first section of core that has 85% Total Core Recovery of Grade III or better, over a length of greater than 5 m. Bands of up to 100 mm of Grade IV and V material or core loss are generally acceptable within this core length. Although this approach to determine the depth of rockhead is reliable in most situations, it has to be recognized that corestones and core slabs up to 10 m thick can overlie thick zones of highly to completely decomposed rock (see Case Study 11.1).

When interpreting the results from ground investigations, it is vital to take into account the orientations of the joints, faults and other discontinuities that have been recognized across the project site. Just connecting the rockhead depths from adjacent boreholes may give a completely false impression of the ground conditions. This point is well illustrated in a section through weathered granite that was exposed in a quarry in Hong Kong (Figure 11.9). If this section had not been exposed, there would have been over 5 m difference in the depth of the estimated rockhead surface from two hypothetical drilling programmes. Using the shallower rockhead surface would increase the risks of pile failure.

Mass weathering grades

A useful classification of weathered rocks is the use of the rock mass classification that maps surface outcrops with respect to zones defined by relative percentage of rock versus soil. Four zones are identified between fresh rock and residual soil, where PW stands for the range of rock percentages within partially weathered material: PW 0/30, PW 30/50, PW 50/90 and PW 90/100. Figure 11.10a graphically displays the mass weathering grades (after Geotechnical Control Office, 1988). Such classification allows a mass weathering grade map to be produced and a ground model to be developed in which the zones PW 0/30 and PW 30/50 act as soils with the rock components having no influence on the mass parameters. A cut slope in granite from northern Spain displays the mass weathering grade zones (Figure 11.10b), whose boundaries are largely defined by the different joint sets.

WEATHERING CHARACTERISTICS

In tropical areas with high humidity and rainfall, the weathering process is more rapid and attains greater depths, whereas in more temperate climates, the weathering is less penetrative and disintegration of the rock mass is dominated by mechanical processes, although the chemical breakdown of the constituent minerals can be significant. In recently glaciated regions, the weathering process is generally not as intense due to lower temperatures, humidity and rainfall and of course the short length of time that the rocks have been subjected to the atmosphere.

(a)

(b)

☐ Colluvium

▨ Grades IV and V
granite

▧ Grades I to III
granite

╱ Rockhead from
blue boreholes

╱ Rockhead from
green boreholes

10 m

(c)

Figure 11.9 Depth of weathering exhumed in a quarry face that is controlled by tectonic joints, sheeting joints and a fault. (a) View of a face in Mount Butler Quarry, Hong Kong. (b) Sketch of the soil–rock distribution in the face. Note that the rockhead surface is nearly vertical in the central part of the section and shallowly dipping elsewhere. Two rockhead surfaces are shown that could have been drawn from two different hypothetical sets of boreholes. (c) Detail of narrow Grade IV weathered zone adjacent to fault surface in Grade III granite.

(a)   (b)

Figure 11.10 Mass partial weathering grades. (a) Idealized column through weathered profile with grades identified. (b) Section of a road cut mapped by using partial weathering grades, northern Spain.

The weathering characteristics of the main rock groups are described below, but in many cases, it is not the composition of the rocks that is most important regarding their propensity to weather, but rather the discontinuities that cut them, for example joints, foliations, shear zones and faults.

Plutonic rocks

The weathering grades for granites have been described, and examples are given in the 'Weathering grades' section. Figure 11.11 presents a typical run of weathered medium-grained granite that varies from Grade I to Grade IV over a vertical length of less than 3 m, with the most intense weathering being associated with low-dipping sheeting joints. It is the weathering along joints that, in general, determines the overall strength of homogeneous plutonic rock masses. This is well illustrated by the rock columns extracted as part of the excavation for bored piles in Hong Kong, which have failed along horizontal weak zones of Grade III material defined by sheeting joints (Figure 11.12).

Weathering of other plutonic rock types can be regarded as similar; however, rocks with little or no quartz may weather more rapidly and to greater depths than granitic rocks, due

Figure 11.11 Weathering of medium-grained granite seen in a vertical site investigation borehole, Hong Kong. Length of core box is 1 m. Weathering varies from Grade I to Grade III with narrow Grade IV zones associated with shallowly dipping joints.

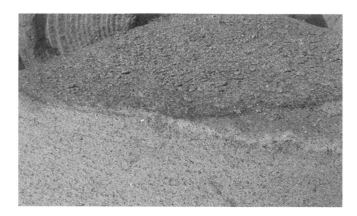

Figure 11.12 Medium-grained, homogeneous granite columns from a bored pile excavation, Hong Kong. The fresh granite (Grade I) displays no joints and has a sharp contact with the overlying weathered rock (Grade III). The columns have broken along a near horizontal seam of Grade III material within a sheeting joint (see also Figure 5.4). View width 2 m.

to the higher percentages of feldspar and ferromagnesian minerals. However, their weathering characteristics are almost identical to granitic rocks, as illustrated by corestone development in syenites from Portugal (Figure 11.8c and d). Plutonic rocks with greatest percentage of ferromagnesian minerals, such as dunite which consists mainly of olivine and plagioclase feldspar, weather to iron-rich soils.

Grain size can also control the degree of weathering in that the finer-grained plutonic rocks are commonly more resistant. Such original variations in the plutonic rock types can control their overall weathering patterns and therefore the topographic and photogeological expressions of different rock compositions (see Figure 5.12a). Primary plutonic fabrics, such as compositional layering or mineral alignments, may have local control on the weathering patterns, but they are of only minor importance.

Figure 11.13 illustrates a typical weathering profile across hilly ground. Here, as has been emphasized previously, it is the discontinuities, including sheeting joints, tectonic joints and faults that control and dominate the weathering patterns and depths of weathering.

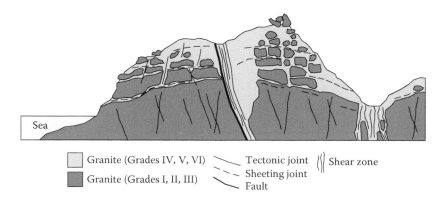

Figure 11.13 Schematic cross-section of a typical weathering profile in granite displaying the controls on the rock decomposition and the depth of weathering.

Volcanic rocks

The weathering of volcanic rocks is extremely varied, which is a reflection of their wide range of composition from quartz-free basalt to quartz-rich rhyolite and their different modes of emplacement, for example as lavas or pyroclastic flows. Features such as bedding, cooling joints and internal fabrics (e.g. flow banding, welding foliations, vesicular structures) will also influence the weathering characteristics of a particular volcanic rock. In addition, the age of the volcanic rock can sometimes be paramount in the development of their weathering profiles. Volcanic rocks from recent eruptions can weather extremely rapidly, in contrast with geologically ancient volcanic rocks, such as welded ash-flow tuffs, which can be extremely resistant to weathering. As a result, it is impossible to cover all situations, and in many cases, a local classification of the weathering grades needs to be established. Examples of the controls of weathering, which can be applied to a wide variety of different volcanic rock types, are described below.

The weathering of the Mesozoic pyroclastic rocks of Hong Kong follows the classification as that established for the plutonic rocks described earlier. The criteria used to distinguish Grade I (fresh volcanic rock) to Grade VI (completely decomposed soil) have been found to be largely equivalent and therefore a useful tool for an estimation of the geotechnical characteristics. Figure 11.14 displays the range of weathering grades from fresh rock (Grade I) to moderately weathered material (Grade III) in a fine-ash, vitric tuff in ground investigation drill core from Hong Kong. Here, the weathering of this essentially homogeneous rock is controlled by the joints with the most intense decomposition being associated with low-dipping sheeting joints.

Weathering of layered pyroclastic tuffs in a temperate climate is well illustrated by a series of cut slopes in Turkey where there is gradation from slightly decomposed rock exposed in the lowest cut slope through moderately weathered rock in the middle slope to highly weathered rock in the upper slope. There is a distinct boundary between the iron-stained Grade IV material and the more weathered rock. This boundary generally follows the layering at approximately 20 m below the original ground surface but is depressed adjacent to some joints and minor faults (Figure 11.15a). The dominance of joint-controlled weathering is apparent in an outcrop of layered pyroclastic rocks from Hong Kong (Figure 11.15b) where the zonation of weathering grades spread out from the different joint sets, although there is a very slight increase of iron-stained weathering along some of the thin layers. Weathering can vary markedly over short distances as a result of localized jointing and faulting, for example the weathering of a massive tuff in Hong Kong beneath colluvium deposits displays very varied weathering patterns over short distances, which have been controlled by the ease of

Figure 11.14 Vertical site investigation borehole through fine-ash, vitric tuff, Hong Kong. Weathering varies from fresh rock (Grade I) to iron-stained rock adjacent to joints (Grade II) and broken material (Grade III) in low-dipping sheeting joints. Length of core box 1 m.

(a)

(b)

(c)

Figure 11.15 Weathering of volcanic rocks. (a) Cut slopes through a sequence of Quaternary layered pyroclastic tuffs. The weathering extends down to 20 m below the original ground surface, Turkey.
(b) Joint-controlled weathering in a Jurassic layered pyroclastic tuff, Hong Kong. (c) Weathering,
iron staining and disintegration of recent volcanic breccia, Mount Etna, Sicily.

infiltration of groundwater (Figure 11.15c). The possibility of lateral changes in weathering
grades should be appreciated when interpreting borehole data, which only provide a very
limited sample of the rock/soil conditions.

Weathering of recent lava flows displays a rapid oxidation of the flow surfaces on exposure to the atmosphere and the establishment of vegetation, for example the basaltic flows
on Mount Etna, Sicily, erupted less than 20 years ago display a thick iron-stained weathering crust (Figure 11.15c).

Sedimentary rocks

The weathering of sedimentary rocks is mainly dependent on grain size, composition of
the constituent grains, nature of the matrix cement and the rock type of the clasts. It is
impossible, therefore, to present the weathering characteristics of a complete range of sedimentary rocks; rather focus is placed on the general controlling factors that will affect the
geotechnical parameters of such rocks. A universal definitive weathering classification for
all types of sedimentary rocks is not available and could be very misleading if specific engineering properties are assigned to particular weathering grades. For example, the weathering classification for chalk will use different criteria than those for a quartz sandstone.

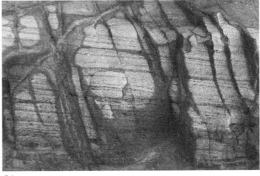

(a)   (b)

Figure 11.16 Weathering of sedimentary rocks. (a) Weathering of a horizontal, thinly bedded sequence of siltstone and fine-grained sandstone. Each layer weathers slightly differently, but the overall weathering pattern is controlled predominantly by sub-vertical joints, Peng Chau, Hong Kong. (b) Iron staining of thinly bedded sandstone along joints and to a much lesser extent along specific thin beds, Tolo Channel, Hong Kong.

Each sedimentary rock sequence must be evaluated separately and the different weathering characteristics identified and geotechnical attributes determined for each situation. As an illustration, the sequence of mudstones and siltstones exposed in an open-pit gold mine in Malaysia (Figure 11.1) shows a wide variability in weathering characteristics of the different sedimentary strata, although the intensity of weathering overall is controlled by the depth beneath the original ground surface. However, in many sedimentary sequences, the different compositions of the individual layers will be highlighted by weathering, but other discontinuities, such as joints, will strongly influence the overall susceptibility to weather. The bedding in a mudstone and siltstone sequence from Hong Kong is picked out by weathering, but weathering along joints is more prominent (Figure 11.16a). Similarly, in a finely bedded tuffaceous sandstone sequence, only a few thin beds are selectively weathered (Figure 11.16b). In summary, weathering effects can change abruptly across an area or even an outcrop composed of bedded sedimentary rocks, so care has to taken when assessing the intensity and distribution of the weathering.

The composition of the constituent grains and the cements is the most important factor in the determination of the susceptibility of a particular sedimentary rock to weather. Simplistically, rocks rich in resistant minerals such as quartz (sandstone) are less readily weathered than those rich in easily altered mica and clay minerals (mudstone). However, where the cement binding of the sedimentary grains is quartz (Figure 4.5a), the rock will be extremely hard, whereas if the cement is calcite (Figure 4.5b), it will be much more readily weathered. In addition, many limestones are composed partially or totally of broken shells, and as a result, their weathering is largely controlled by the dissolution of the carbonates by groundwater. The resulting landform is called karst that includes features such as caverns, collapse structures and rock pinnacles (see Chapter 12).

In fine-grained rocks, for example mudstone, siltstone and fine-grained sandstone, there is a wide compositional spectrum of component grains, which may vary from quartz and feldspar grains and detrital clays to a suite of heavy minerals (e.g. magnetite and zircon). Of these, feldspar is most prone to weathering, and therefore its percentage will largely determine the susceptibility of the rock to weathering and the loss of strength. The composition of the matrix may vary from quartz to calcite which again will affect the weathering characteristics of the rock. As a consequence of all these variables, it is impossible to present the complete range of

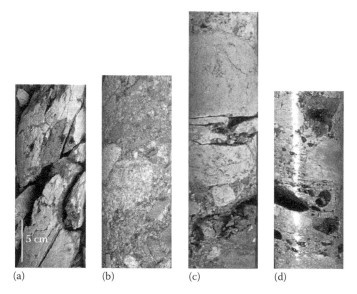

Figure 11.17 Weathering of sedimentary rocks from vertical site investigation boreholes, Hong Kong. (a) Moderately to highly decomposed, steeply dipping siltstone with preferential decomposition along the bedding. (b) Partially decomposed sedimentary breccia displaying different intensities of weathering of granite and siltstone clasts and the matrix. (c) Extensively decomposed matrix to sedimentary breccia in which the fine-grained granite clasts are relatively unweathered. (d) Decomposition and solution of calcareous clasts in a slightly metamorphosed, sedimentary breccia.

the effects of weathering that may be imposed on sedimentary rocks. A core sample from a site investigation in Hong Kong (Figure 11.17a) illustrates the weathering of a siltstone, where the intensity of decomposition is, in part, controlled by slight compositional differences in the rock.

In coarse-grained sedimentary rocks, including some sandstones, conglomerates and sedimentary breccias, the weathering characteristics are controlled by the composition of the constituent grains and clasts and the nature of the cement matrix. In conglomerates and breccias, the intensity of weathering of the different clasts will vary, for example in a core sample of breccia from Hong Kong, the granite clasts have weathered more intensely than the adjacent siltstone clasts (Figure 11.17b). However, in other breccias, it may be that the matrix is more weathered in comparison to the clasts. In the core sample shown in Figure 11.17c, the fine-grained granite clasts are only slightly weathered, whereas the matrix is iron-stained and completely decomposed. An exceptional illustration of the different weathering characteristics of clasts and matrix is where calcareous clasts have been almost totally weathered out, leaving numerous small cavities within a relatively strong matrix (Figure 11.17d). Here it is the percentage of former clasts that will determine the geotechnical properties of the rock mass.

Metamorphic rocks

In general, the weathering of metamorphic rocks is dependent on their composition, grade of metamorphism and the intensity and nature of the metamorphic/tectonic fabric. Those rocks rich in easily decomposed minerals such as feldspar, clay minerals, micas and carbonates weather much more readily and to a greater depth than those rich in more resistant minerals, in particular quartz. The weathering of low- and medium-grade metamorphic rocks, including slates, phyllites and schists, is a reflection of the ease of penetration by groundwater, which is largely controlled by tectonic fabrics such as schistosity, although in some localities jointing can be of more importance. In areas of steeply dipping, closely spaced foliation, the weathering is more significant than in areas where the foliation is sub-horizontal and less intense. In higher-grade rocks, including granitic gneisses and migmatites, the metamorphic foliation is in general not so significant with respect to their propensity to weather; of more importance is the relative percentage of the easily decomposed minerals. Commonly, gneisses weather in a similar manner to granitic rocks.

Fault rocks

Faults commonly are conduits for groundwater and therefore, in general, are highly susceptible to the weathering process. Increased jointing and sheared fabrics, within and close to fault zones, make faults very susceptible to weathering, erosion and the development of negative topographic features. In fault zones, brittle and soft materials are commonly juxtaposed and therefore are fragmented by the drilling process or not recovered at all. This can be a major shortcoming as Mazier samples of this material can provide undisturbed intact samples through a fault zone that can give essential information on the style and intensity of the fault movements. For example, Figure 11.18 presents a selection of different styles of faulting recovered in Mazier samples: a mylonite from ductile thrust fault, a breccia

(a)   (b)   (c)

Figure 11.18 Weathering of fault rocks displayed in Mazier samples, Hong Kong. All samples are completely decomposed and very soft. (a) Relict mylonitic foliation in a ductile fault. (b) Fault breccia in brittle fault with preservation of original structures in both the matrix and clasts. (c) Shear zone in foliated granodiorite cut by later quartz vein.

from a brittle normal fault and foliated granodiorite from a shear zone. This information on fault dynamics would have been totally lost if they not have been recovered intact. The importance of recovering completely decomposed fault material in Mazier samples is well illustrated in Case Study 11.1, which describes the foundation conditions beneath a school in Hong Kong. Here, a major, moderately dipping, shear zone controlled the depth of weathering and the form of the rockhead surface.

CLAYS IN THE WEATHERING PROFILE

The breakdown of feldspar and mica under tropical conditions to a variety of clay minerals, particularly kaolin, greatly changes the geotechnical properties of soils within weathering profile (Geological Society, 1997). The clay is mobilized in circulating groundwater and precipitated in joints or as clay-rich layers close to the ground surface. In places the clay is associated with black, hydrated manganese oxide minerals (Figure 11.19a). The formation of kaolin under these conditions is distinct from the kaolin found in hydrothermal

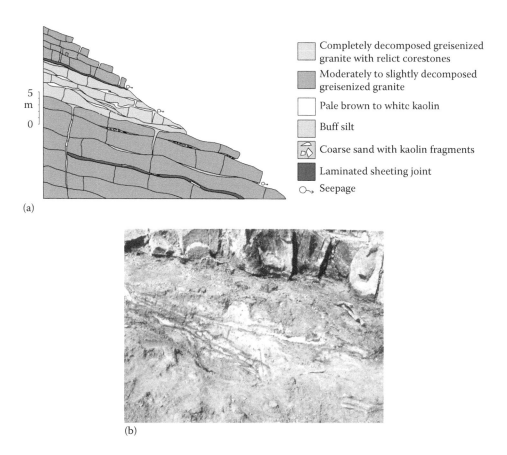

(a)

(b)

Figure 11.19 Cross-section of cut slope in granite above new development, Eastern New Territories, Hong Kong. (a) Schematic section of site showing different occurrences of kaolin in granite. (After Fletcher, 2004.) (b) Kaolin seams within completely decomposed granite overlain by a slab of rock.

alteration zones in granite, although it is often difficult to distinguish their origins. For example, the deeply kaolinitized granites that form the 'badland' topography in Hong Kong (Figure 11.1b) could be argued to have been formed by one of these processes or a combination of both. The clays are grouped under the generalized term 'kaolin', although this will include such minerals as kaolinite (*sensu stricto*), smectite, gibbsite, halloysite and other clay minerals, all of which have the basic composition of hydrated aluminosilicates, with some replacement by Ca, Fe and Mg. Kaolinite, in particular, is important as it has a platy structure with a low coefficient of inter-particle friction, and the crystals may become orientated and form continuous shear surfaces. Soils with over 40% clay fraction have properties that are defined by the presence of the clay (Geological Society, 1997). Kaolin development in weathered rocks, therefore, can be a very important factor in slope stability investigations in tropical climates, and in Hong Kong, several recent landslides have been attributed to the presence of adversely orientated thin seams of kaolin in the soil profile (see Case Study 6.1).

Kaolin has a friction angle of less than 18° in contrast with on average 26° in completely decomposed granite or volcanic rocks. Such lowering of the friction angle and the impermeable nature of the kaolin seams can result in the failure of natural hillsides or cut slopes. The most significant kaolin seams have been traced for tens of metres along strike and some have thicknesses of over 20 mm, although even thin seams, less than a few millimetres thick, have been known to act as basal rupture surfaces. A study of a cut slope in Hong Kong (Figure 11.19) illustrates the features that are associated with the development of kaolin in the weathering profile of a medium-grained granite that is intersected by subvertical tectonic joints and a set of low-dipping sheeting joints. The cut slope lies approximately 10 m below the original ground surface. Kaolin is found predominantly within a zone of highly to completely decomposed material (Grade IV/V) that lies below a slab of rock (Grade II/III) (Figure 11.19a). Seepage of groundwater occurs close to the base of the soil horizon and above significant kaolin seams, thereby forming a perched water table. Several white kaolin seams and lenses follow relict sheeting joints and attain thicknesses of 250 mm in places (Figure 11.19b). Some seams can be traced for over 20 m along the strike of the weathered zone. Kaolin is also found within joints below rockhead, both in the sub-vertical joints and sheeting joints as angular fragments set in a silt matrix, presumably washed in by recent groundwater circulation. The adverse orientation of the kaolin seams with respect to the cut slope would suggest that failure could occur in the overlying rock and soil.

Identification of thin kaolin seams from ground investigation boreholes can only be made by careful sampling techniques and subsequent study of the soil samples in opened Mazier tubes or SPT liners. These samples, which are commonly not opened, can provide vital data on the distribution and orientation of kaolin seams in the weathering profile, thereby providing essential data on the geotechnical properties of soil mass.

Swelling clays

In some volcanic rocks such as basalt and the more silica-rich, ash-flow tuffs, the mineral montmorillonite can be formed within weathering profiles. Montmorillonite can absorb water within the crystal lattice under wet conditions, thereby swelling and radically altering the geotechnical properties of the soil, in particular the creation of very low shear strengths. On exposure to dry atmospheric conditions, the montmorillonite shrinks and cracks very rapidly, which can cause active changes to the ground engineering works, for example in recently excavated cut slopes (Figure 11.20).

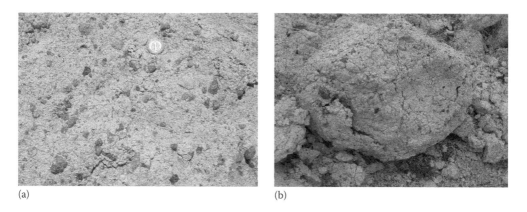

(a)   (b)

Figure 11.20 Effects of swelling clays in tuff. (a) Fresh tuff immediately after exposure during excavation of cut slope. (b) Tuff after 3 months of exposure to the atmosphere. The boulders display discolouration and irregular expansion cracks.

CASE STUDY 11.1   DEEP WEATHERING ZONES IN GRANODIORITE: SCHOOL FOUNDATIONS, TIN SHUI WAI, HONG KONG

[22°27.8′N 114°0.4′E]

Weathering of a granodiorite body at a development site for a school in the Northwest New Territories of Hong Kong (Figure CS11.1.1) illustrates the difficulties that can arise in the determination of founding levels in areas of complex rockhead surfaces (Fletcher et al., 2003). Following the site investigation and development of the geological model, the footprints of the school buildings, as originally planned, were relocated in order to give low-risk and cost-effective pile foundations. The geological model was dependent on the opening of all the Mazier samples, thereby extending the geological line work into the soil rather than stopping it at rockhead – a

Figure CS11.1.1 Tin Shui Wai site showing a previously completed school and high-rise buildings.

common practice. It also illustrates a situation where thick sections of completely weathered rock are present below assumed rockhead, as defined by 5 m of Grade III or better rock.

GEOLOGY

The site is unlain by a 100 m thick tabular body of granodiorite that is flanked by a sequence of metasedimentary rocks (Figure CS11.1.2). The contacts of the granodiorite are defined by thrust faults that dip at moderate angles towards the NW. The central part of the granodiorite body is medium grained and equigranular (Figure CS11.1.3c), whereas close to the thrusts, it has been intensively sheared and displays a penetrative foliation (Figure CS11.1.3d).

The sedimentary rocks consist of interbedded siltstone, sandstone and marble that have been dynamically metamorphosed to a low grade. Minor folds and slaty cleavage are present within the finer-grained lithologies (Figure CS11.1.3b) and the marble layers display transposed bedding and shear structures (Figure CS11.1.3a). The average orientation of the compositional layering of the metasedimentary rocks and all the tectonic foliations dip at approximately 45° towards the northwest. All these structures indicated the presence of intense shear zones that would greatly affect the geotectonal properties of the rock and their weathering characteristics.

Moderately decomposed to fresh rock underlies most of the site at depths ranging from −45 to −55 m below ground level and therefore provides suitable founding levels for bored piles. In the central part of the site, the rockhead is depressed to over 90 m below ground level.

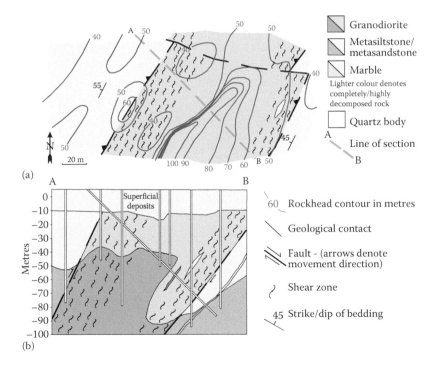

Figure CS11.1.2 Geology of the Tin Shui Wai school site. (a) Geological map with rockhead contours. (b) Cross-section with representative boreholes.

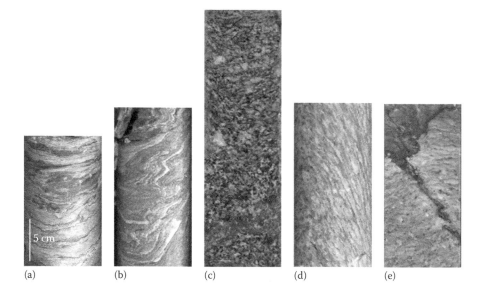

Figure CS11.1.3 Borehole samples from site investigation. (a) Marble displaying penetrative shear foliation and disruption of the original sedimentary bedding. (b) Recumbent, small-scale folds in schist. (c) Undeformed, completely decomposed granodiorite – Mazier sample. (d) Strongly foliated granodiorite. (e) Completely decomposed, foliated granodiorite – Mazier sample.

Here, a 30 m thick zone of completely decomposed, sheared granodiorite (Figure CS11.1.3e), containing large lenses of deformed quartz, dips beneath a slab of moderately decomposed to fresh rock. This zone of soil was probably the result of increased groundwater flow above and adjacent to one of the thrust planes.

ENGINEERING CONSIDERATIONS

- Steep to overhanging depression in rockhead caused by decomposition along a shear zone.
- Over part of the site, a wedge of rock is underlain by 30 m of soil, thereby making it unsuitable for founding bored piles.
- Strong lenses of quartz occur within the decomposed shear zones – these could have damaged driven piles.
- Cross-sections based only on joining up geological contacts between boreholes can be very misleading.
- Location of buildings changed on account of subsurface geological conditions.

CASE STUDY 11.2   DESERT WEATHERING AND SUPERFICIAL DEPOSITS: RAILWAY ALIGNMENT, GOBI DESERT, MONGOLIA

[43°5.3′N 106°40.1′E]

N. R. Wightman *Aquaterra Consultants*
and
A.D. Mackay
Nishimatsu Construction

The preliminary investigation for a proposed 250 km railway from a coal mine in the southern part of the Gobi Desert to the border with the People's Republic of China, included a desk study and terrain evaluation followed by a walk over survey and extensive ground investigations (Mackay et al., 2013). The alignment descends from 1500 m above sea level in the north to around 900 m in a south-central section. Engineering risks were estimated by consideration of the physical characteristics of the ground as a sum of the geological, geomorphological and recent environmental history. In particular, the risks associated embankment placement on a variety of recent deposits and the weakening effects of different rock types were investigated. Most of the proposed railway alignment crosses over expansive plains covered by superficial alluvial, lacustrine, colluvial and aeolian deposits (Figure CS11.2.1). These flat areas are flanked by ranges of low hills, topped with sharp rock ridges. Cementation and disturbance of the superficial deposits resulting from recent changes in climate leading to the present-day cold desert environment have, in places, markedly altered the properties of these deposits and underlying bedrock.

GEOLOGY

The published geological map of the area (Figure CS11.2.2b) provides a broad indication of expected rock types to be predicted in exposures but provides little information as to their distribution below the superficial deposits. The satellite image of the area (Figure CS11.2.2a) clearly defines the craggy darker areas in the southern part of the alignment, where volcanic and granite rocks are exposed. Elsewhere, the more uniform, lighter-coloured, topographically subdued areas are blanketed by superficial deposits. A drilling programme has now revealed

Figure CS11.2.1 Surface features of the railway alignment. Weakly cemented duricrust overlying loose sand in aeolian soils and mini dunes.

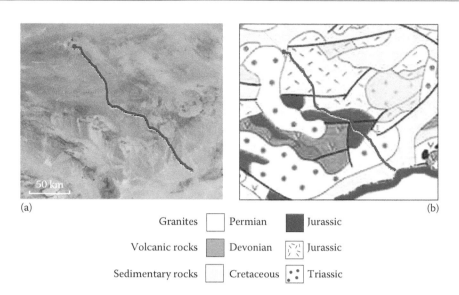

(a)

(b)

Granites	☐	Permian	■	Jurassic
Volcanic rocks	▨	Devonian	▨	Jurassic
Sedimentary rocks	☐	Cretaceous	⠿	Triassic

Figure CS11.2.2 Area of the planned railway alignment (marked by red line). (a) Google Earth, Landsat image of topographic features from the Tavan Tolgoi Coal mine in the north to the border with China in the south. (b) Geological map of the alignment area. (From Mongolian Geological Survey, 1:5,000,000-scale Geological Map of Mongolia, 2010.)

(a)   (b)

Figure CS11.2.3 Rock outcrops. (a) Wind-eroded conglomerate. (b) Hard, jointed basalt lava on a hill top ridge.

the bedrock geology beneath the superficial deposits and determined their state of decomposition. The bedrock of the area consists mainly of sedimentary rocks, including conglomerate, sandstone and siltstone, with seams of coal, which are exposed in a few localities in the northern part of the area; the conglomerates, in particular, can form upstanding outcrops (Figure CS11.2.3a). In the south, volcanic rocks and subvolcanic intrusions form outcrops across the higher ground (Figure CS11.2.3b). Superficial deposits have accumulated in a variety of geomorphological environments across the area and continue to accumulate today. They include colluvial sand and silt adjacent to the higher ground, alluvial sand and gravel in ephemeral streams and rivers, lacustrine silt and evaporite deposits in sporadically flooded topographic depressions and ubiquitous aeolian sand dunes and sheets. Weathering in this harsh, cold environment has resulted in the rock substrate being deformed by frost action with the formation of frost wedges

(a)

(b)

Figure CS11.2.4 Near-surface ground features. (a) Two metre deep ice wedge in weathered conglomerate filled with sand. (b) Cutting in cryoturbated residual soil of mudstone with
evaporitic concentrations of white calcite as layers and in veins at depth.

in the hard lithologies (Figure CS11.2.4a) and cryoturbation and evaporite deposition in softer
residual soils (Figure CS11.2.4b). Recent cementation of the superficial deposits has resulted in
the formation of iron-rich duricrust over many of the older sand dunes (Figure CS11.2.1).

ENGINEERING CONSIDERATIONS

- Superficial deposits of varying composition and strength overlie a wide variety of sedi-
  mentary rocks.
- Exposed volcanic rocks and igneous intrusions are suitable for railway ballast.
- Aeolian soils have a weak duricrust covering loose fine deposits causing weak embank-
  ment foundation conditions.
- Cryoturbation and evaporite concentration have changed the soil compositions and their
  geotechnical properties.
- Summer storm runoff causes surface channelization and deposition of alluvial gravely soils.

Chapter 12

Karst

Chemical weathering of limestone and other carbonate-rich rocks results in the formation of *karst* – a landform developed through the solution of the carbonate minerals by rainwater enriched in carbon dioxide both at ground surface and along discontinuities, geological contacts and bedding planes at depth. Karst features include caverns, subterranean rivers, sub-circular collapse structures (swallow holes), rock pinnacles (Figure 12.1) and dry river valleys. In older karst systems, these features are commonly filled by more recent sediments, including washed-in sands and silts, sedimentary breccias formed of angular blocks of limestone fallen from the cavern roofs and chemical deposits of calcium carbonate called *tufa*. Karst is most well developed in tropical and subtropical areas where humid conditions prevail. In Hong Kong, the effects of karstification extend to over 200 m below present sea level (see Case Study 12.2). In many areas, old karst surfaces are covered by alluvium or thick deposits of soil consisting of the insoluble residue of the weathered carbonate rocks.

One of the largest cities that has had to overcome the challenges of construction on limestone and marble is Kuala Lumpur, Malaysia (Figure 12.1), where about a third of the metropolitan area is underlain by karst (Mitchell, 1986). The calcareous rocks are confined to a large lens up to 1850 m thick that is enveloped by quartz/mica schist, graphitic schist and quartzite (Hutchinson, 1989). This folded metamorphic sequence is intruded by Mesozoic granites, crosscut by a series of faults and mineralized veins and covered by alluvial deposits of clay, sand, gravel and peat. Karst features are characterized by highly irregular bedrock morphology which includes rock pinnacles with vertical to overhanging walls up to 60 m in height, isolated rock slabs, deep channels along contacts with granite, sinkholes, sloping rock surfaces, slump breccias and cavities (Yeap, 1986). In Hong Kong, there are a variety of geological settings that control the distribution, intensity and depth of the karst development, and these are schematically illustrated in Figure 12.2.

ENGINEERING CONSIDERATIONS

Karst includes some of the most challenging geological phenomena to be encountered by the site engineer as voids, extreme morphological variability of the rockhead surface, soft cavity-fill material, corestones in soil and the possibility of on-going solution and collapse can all be encountered. Ground investigation in karst areas have to be extensive and may require specialist drilling techniques to recover the softer materials. Surface geophysical surveys, including microgravity, resistivity and seismic methods can provide vital information on the subsurface variability, in particular the identification of cavities in rock, whereas downhole geophysics can help in determining the 3D picture of the rock/soil interface. It is not uncommon for the karst surfaces and other features to be blanketed by

Figure 12.1 Karst towers in a drowned coastal landscape, Phuket, Thailand.

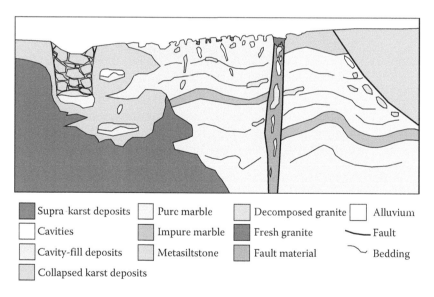

■ Supra-karst deposits	☐ Pure marble	▨ Decomposed granite	☐ Alluvium
☐ Cavities	▨ Impure marble	▨ Fresh granite	⎯ Fault
☐ Cavity-fill deposits	▨ Metasiltstone	▨ Fault material	⌣ Bedding
▨ Collapsed karst deposits			

Figure 12.2 Schematic diagram of the geological settings of the karst features, based on the experience in Hong Kong.

superficial deposits, for example in Kuala Lumpur, the alluvium is up to 20 m deep (Yeap, 1986). Thus, although alluvial plains may seem ideal sites for the construction of new developments (see Case Study 12.1), they may hide complex ground conditions at depth, and any desk study must consider the possibility of the presence of unexposed limestone and marble.

One of the key engineering issues of foundation design in karst regions is the nature of the subsurface bedrock morphology and whether voids occur below this surface and, if so, whether they are filled with sediment. The upper surface can be mapped using borehole

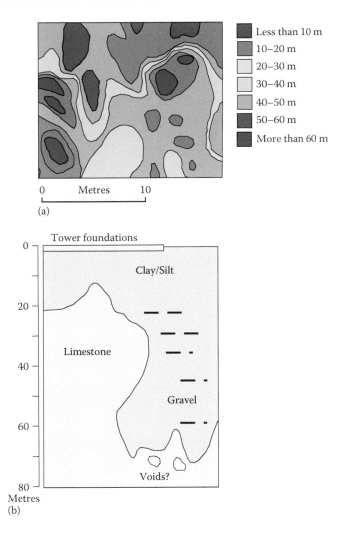

Figure 12.3 Karst in Kuala Lumpur, Malaysia. (a) Typical contour map of the depth to bedrock surface beneath a building footprint. (b) Cross-section of ground conditions beneath the Pan Pacific Hotel Tower. Ironstone occurs as lenticular layers and boulders within superficial deposits. (After Mitchell, J.M., Foundations for the Pan Pacific Hotel on pinnacled and cavernous limestone, in: Chan, S.F., 1986.)

intersection data to produce a rockhead contour map (e.g. Figure 12.3a), but even with close spacing of boreholes, unexpected 'cliffs' or 'overhangs' in the limestone/marble can be missed (Figure 12.3b). Microgravity surveys have been used to provide an overview of the rockhead surface, but they commonly lack the detail required for foundation design. To determine the exact location and size of voids and the composition of the cavity-fill material, if present, beneath the rockhead surface, which is extremely difficult, the engineer should be aware that voids have been 'identified' on borehole logs by the lack of core recovery or a sudden drop of the rods during drilling. However, using standard drilling procedures, soft cavity-fill material in limestone would react in a similar manner. Down-hole geophysical investigations may help to locate voids close to a borehole, and specialist drilling techniques, such as the use of polymer drilling fluids, can assist in the recovery of soft materials at depth. A classification system for karstic marble has been devised for the assessment and determination of the sound founding conditions in the karst areas of Hong Kong (Chan and Pun, 1994). It is based on the fracture state of the rock and the core recovery ratios, thereby providing a range of conditions for the rock mass from very good quality, through rocks moderately affected by dissolution, to rocks with substantial voids.

Many of the buildings in Kuala Lumpur, Malaysia, have required extensive ground investigations to determine the subsurface conditions. The shape and depth of the bedrock surface is often highly variable, for example on some sites the depth of the surface ranges from less than 10 to more than 60 m over a horizontal distance of less than 5 m (Figure 12.3a). To avoid the most extreme bedrock morphologies, it has been the practice, where possible, to reposition or redesign the buildings. This was the case with Kuala Lumpur's highest building, the twin Petronas Towers, which was moved some 60 m from its originally planned location to avoid a 'cliff' in the bedrock surface. The architectural plans of the Pan Pacific Hotel were modified so that the foundations of the bedroom towers were not affected by a 50 m high overhang in the bedrock surface (Figure 12.3b) (Mitchell, 1986). Overall, such ground conditions pose many geotechnical problems and require detailed subsurface ground investigations and foundation designs that incorporate bored piles to depths of over 60 m, driven H-piles, grouting, minipiles, barrettes and rafts.

In summary, some of the geotechnical factors that should be considered by the engineer when assessing a project site in a karstic area are as follows:

- The uncertainty of the bedrock morphology – the rockhead can unexpectedly become vertical over a lateral distance of a few metres.
- Cavities may occur at depth as part of subsurface river channels and cave systems.
- Upward migration of cavities from below may cause catastrophic collapse of shallow foundations.
- Proximity to faults, highly fractured rock and contacts with other rock types, the karst features may extend to great depths.

Two case studies are included in this chapter, both of which are located in Hong Kong where no limestone/marble is exposed at surface and was only first encountered in subsurface site investigations in the 1980s. The first case study concerns Yuen Long Railway Station (Case Study 12.1) and describes the extent and variability of karst features related to faults and contacts with igneous intrusions. The second case study describes the complex geology beneath the foundations for high-rise apartment blocks at Tung Chung (Case Study 12.2). There the karst features, including thick cavity-fill deposits to a depth of –120 mOD, are confined to a former exotic block of marble enveloped in completely decomposed granite.

KARST MORPHOLOGY

All limestones and marbles dissolve in acidic rainwater to produce a wide variety of surface and subsurface geomorphological features, including rock pinnacles (Figure 12.4a), towers (Figure 12.1), solution collapse holes (Figure 12.4d), incised limestone pavements and underground cave and river networks. Many of these karst features continue to evolve today, but some are essentially inert and formed during ancient weathering episodes. An example of an inactive cave system is exposed in a cliff section near in Phuket, Thailand. Here, a series of caves are distributed along faults and certain bedding surfaces high above present sea level (Figure 12.4b). Some of the caves are partially filled with sediment and most

(a)

(b)

(c)

(d)

Figure 12.4 Examples of morphological features in areas of karst. (a) Small pinnacles of limestone exposed by the removal of overlying alluvium during previous tin mining operations, Malaysia. (b) Old cave system exposed in a cliff section, Thailand. (c) Swallow hole with vertical limestone sides, Thailand. (d) Entrance to major cave and river system developed along fault. Partial collapse of the rocks above the voids has occurred along the line of a fault, Malaysia.

have stalagmites that have ceased to grow. In the Krabi region of Thailand, the karstification of a former limestone platform (Figure 12.5) started some 2 million years ago. Subsequently, the mature karst landscape was exposed by erosion during uplift and then flooded during sea level rise at the end of the last Ice Age. Many other ancient karst landscapes have either been buried by younger sediments or exhumed by tectonic uplift and erosion, these are referred to as *palaeokarsts*. Karst systems have been identified in rock strata from many geological periods, and these provide evidence for ancient climatic conditions and landscape evolution patterns. For example, a sea cliff section in Carboniferous limestone at Red Warf Bay, Wales, has exposed depressions in an ancient karst surface filled with deltaic sands during a marine transgression which submerged the potholes (Figure 12.6). Recent exposure of the sandstone fill material in these karst structures has resulted in the partial erosion of the material by wave action to form new cavities.

Excellent synopses of the many and varied geomorphological features of karst throughout the world are provided by Jennings (1985) and Sweeting (1972).

Figure 12.5 Flooded ancient karst landscape, near Phuket, Thailand. The near-flat tops of the highest towers are the remnants of an ancient limestone platform.

Figure 12.6 Palaeozoic karst feature. Buff sandstone filling a near circular, swallow hole which formed on a Carboniferous limestone pavement. The lower section of sedimentary rock within the swallow hole has been eroded out by present-day wave action. Red Warf Bay, Wales.

Micro-karst features

The recognition of different solution and decomposition phenomena in boreholes provides an indication of the extent, intensity and possible distribution of karst across a project area, thereby providing information of the possible strength variation of the rock mass and the presence of, and controls on, the formation of cavities. Figure 12.7 illustrates some of the characteristic solution and weathering phenomena that should be assessed in any site investigation, so as to develop the larger-scale geological model. Depending on the composition of the carbonate rock, the acidity of the groundwater and erosive potential of the circulating underground water, the micro-karst features vary from the *in situ* decomposition of the limestone/marble seen as zones of iron-stained and dark microporous material around a

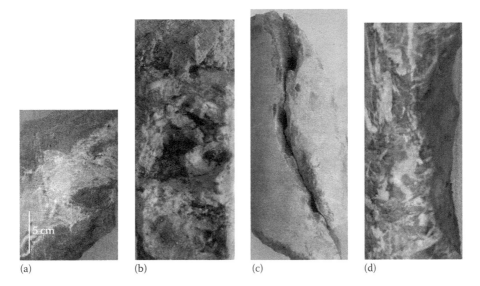

Figure 12.7 Examples of calcareous rocks displaying dissolution and weathering features in borehole core samples, Hong Kong. (a) Iron-stained and microporous weathering zones around small core-stones of limestone. (b) Delicate framework of very delicate weathered marble with cavities. (c) Small cavities developed along a fracture in limestone. (d) Water-worn margin to cavity in veined marble.

core of unaltered rock (Figure 12.7a), or delicate frameworks of the most insoluble carbonate minerals (Figure 12.7b). The effects and extent of karstification can also be recognized by small cavities along solution-widened fractures (Figure 12.7c) and smooth water-worn margins of cavities (Figure 12.7d), where the water flow rates were high.

KARST DEPOSITS

The sediments that are associated with karst development are of two types: those that were deposited in underground cave systems and isolated caverns (cavity-fill deposits) and those that accumulated on top of karstic weathering surfaces (supra-karst deposits). In both environments, hard limestone or marble is juxtaposed against sequences of clay and silt containing cobble- or boulder-sized clasts. It is not uncommon for the karst deposits to be totally or partially eroded away during later ground and surface water action and for alluvial sediments to be washed into the voids or cover the karst surface.

Supra-karst deposits

Supra-karst deposits are the residue from the solution of calcium carbonate and leaching of silicate minerals within limestone bodies, which accumulate in depressions or as a blanket above the sculptured carbonate rock surface. They are commonly referred to as 'Terra Rosa' (red earth) due to their red colouration (Figure 12.8). The thickest deposits primarily develop under tropical to subtropical climates with high annual rainfalls over extended periods of time. Where they occur in arid and temperate climates, they provide an indication of previous climatic conditions. The residual deposits contain abundant iron hydroxide

(a)   (b)

Figure 12.8 Supra-karst deposits. (a) Highly irregular kart surface overlain by red-brown, iron-rich, sandy clays, Ogcheon, South Korea. (b) Red-brown sand and clay concentrated along sub-vertical fractures and cavities. Originally, the limestone would have been covered by a thick layer of supra-karst deposits, Santa Catarina, Portugal.

minerals that readily oxidize to red-brown iron oxides. In places, these deposits can contain fragments of the host limestone, develop ferrous hard grounds or nodules (Figure 12.3b) due to the action of ground water and may be contaminated by the influx of washed-in sediments from the neighbouring areas. The development of a blanket of reddish-brown, supra-karst deposit above a highly irregular karstic surface is exposed in a road cut in South Korea. There, the deposits contain angular limestone fragments that have become detached from the parent limestone during the karst-forming process (Figure 12.8a). Commonly, rainwater will carry materials from the supra-karst deposits downwards into the fractures and voids in the underlying limestone (Figure 12.8b).

Cavity-fill deposits

Active cave systems are formed by the chemical weathering of calcareous rocks by circulating underground water. During this process, caves may become partially filled with washed-in sediments that are deposited on the beds of low-energy streams and the floors of subterranean lakes. As the rate of solution of the limestone/marble diminishes due to the changing level of the water table related to climate change, sea level fluctuations or uplift, the sediments accumulate more readily and the voids become progressively filled; however, the re-establishment of high water flows may erode the cavity-fill deposits. These deposits, therefore, provide a detailed record of past climates, weather patterns and relative sea level changes. For example, in Hong Kong, many of the cavity-fill deposits accumulated during the last glacial period when sea levels were 120 m lower than today.

Cavity-fill deposits include laminated silts that reflect annual variations in rainfall, sands washed into the cavities during high groundwater flows, possibly during storms, and blocks of rock that fell from the roofs and walls of the cavities. Three examples of cavity-fill sequences are described that illustrate the range of deposits. The first is contained within an old cave system that has been dissected and exposed in a cliff section near Phuket, Thailand (Figure 12.9a). The lower part of one cavity was filled with large angular blocks of roof-fall material interbedded with smaller fragments. The upper part of the cavity remained open and the original stalagmites still hang from the roof. All the cavity-fill material has been cemented by calcareous deposits (tufa) that precipitated from

Figure 12.9 Cavity-fill materials. (a) Roof-collapse breccias in partially filled cavity, Phuket, Thailand. (b) Elongate cavity along a fault that has been filled with a bedded sequence of sand, layered breccias and silt, Ha Long Bay, Vietnam. (c) Bedded sequence of mud, silt and fine sand with some layers rich in angular limestone fragments exposed along the walls of an underground river system, Malaysia. Height of section 100 cm. (d) Silty sand with wet sediment deformation structures. Mazier sample. (e) Angular completely decomposed granite fragments in a silty clay matrix overlying laminated silt. SPT liner sample. (f) Faulted, finely laminated sand and silt. Mazier sample. Samples (d–f) from site investigation borehole recovered from below −100 mPD, Tung Chung, Hong Kong (see Case Study 12.2).

percolating groundwater, thereby forming a hard sedimentary breccia. The second example is exposed in a sea cliff in Ha Long Bay, Vietnam, where a series of cavities, which had developed along the line of a fault, have been partially filled with bedded sand and gravel, together with some isolated, small rock fragments (Figure 12.9b). These sediments are water lain except for the inclusion of some roof-fall material. The third cavity-fill deposit is preserved on the sidewalls of an active underground river system in central Malaysia (Figure 12.9c). This well-bedded sequence of clay and silt contains isolated, angular blocks of limestone, which are concentrated in particular layers. It is probable that this sequence was deposited in an underground lake under variable water inflows that reflect the intensity of rainfall at surface.

Figure 12.10 Run of core in marble displaying dissolution, decomposition features and cavity-fill material, using conventional drilling techniques. Diameter of core 8 cm. Sections of fragmented core and/or core loss could equate to cavities, New Territories, Hong Kong.

Recovery of cavity-fill deposits during site investigations, as has been stated previously, is difficult and costly and may require specialist drilling techniques. However, it has been possible to recover undisturbed loose sands, displaying delicate sedimentary structures, in Mazier samples, and sedimentary breccias with completely decomposed granite fragments and finely laminated fine sand, silt and clay in SPT liner samples from depths of over –120 m in Hong Kong (Figure 12.9d through f). A typical run of core using conventional drilling techniques from a project site in karstic marble is shown in Figure 12.10. The limestone displays many karstic features, including solution along some of the joints, decomposition and iron staining of the marble and sections of cavity-fill material composed of brown, slightly layered, silty sand. Core recovery in some sections was very poor and these were referred to as cavities in the core log. However, there is a distinct possibility that they are filled with soft silt or clay that was lost during the drilling process. All these features suggest that karstification is intense at these levels and probably extends laterally some distance.

CASE STUDY 12.1   KARST: RAILWAY STATION FOUNDATIONS, YUEN LONG, HONG KONG

[22°26.85′N 114°1.53′E]

Yuen Long railway station is located on an alluvial plain in the Northwest New Territories of Hong Kong (Figure CS12.1.1). There are no rock outcrops in the immediate vicinity of the site and an anticipated geology relied on the published sub-superficial geological map of the area (GCO, 1989). The geology of the surrounding hillsides, which are composed mainly of volcanic and granitic rocks, provided clues as to the possible regional fault patterns. Marble was only discovered in the area in 1980 when the first site investigation data for the new town became available. By 1987, following several dedicated deep boreholes, it became apparent that large tracts of the area were underlain by marble (Yuen, 1990; Frost, 1992). The solid geological map of the area (Figure CS12.1.2) was constructed mainly from site investigation data and therefore cannot be relied on for site-specific investigations, except where reinterpretation has been possible from more recent borehole data.

Figure CS12.1.1 Yuen Long alluvial plain, which is underlain by karstic marble, meta-siltstone and granodiorite.

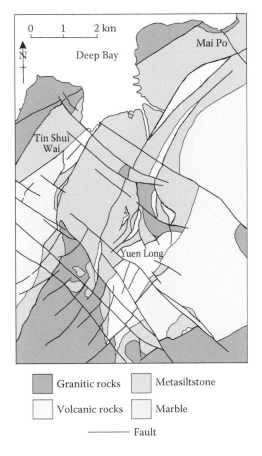

Figure CS12.1.2 Geological map of the Yuen Long region. (After Frost, 1990; Hong Kong Geological Survey, Geotechnical Engineering Office.)

GEOLOGY

The Yuen Long plain is covered by an approximately 10–20 m thick blanket of Pleistocene and recent superficial deposits, including alluvial silts, marine mud and debris flow deposits up to 36 m thick (GCO, 1989). The Carboniferous bedrock consists of a dark grey marble displaying disrupted bedding traces and a more massive white marble unit. A thick sequence of metamorphosed siltstone and sandstone overlies the marble. Several tabular bodies of granodiorite have been intruded into the sedimentary succession. The whole sequence has been deformed into a series of NE-trending folds associated with southeast-directed thrusts that have repeated the succession. A series of northwest-striking, steeply dipping faults also transect the area. Across the region, tropical weathering prior to the sedimentation of the superficial deposits caused significant karst formation and decomposition of the granodiorite. The top of the marble, which lies on average at −10 mPD, was originally sub-planar, but the formation of karst has transformed the local bedrock topography into highly irregular pinnacles and troughs. The solution phenomena extend 10–15 m into the marble with the development of cavities, which for the most part are filled with sediment. Karst features have developed preferentially along and adjacent to fault zones to depths in excess of −120 mPD.

The eastern and central parts of the Yuen Long site are underlain by marble and cut by two, steeply dipping, NW-striking faults; the easternmost consists of a zone of sub-vertical silvers of marble between fault strands and is associated with closely spaced jointing in the marble (Figure CS12.1.3), whereas displacements are confined to a single fault plane. The western part of the site is composed of granodiorite tectonically interleaved with meta-siltstone, both of which have been thrust over the marble. The granodiorite is completely decomposed to depths up to −120 mPD. Figure CS12.1.4 presents a series of split SPT liners, Mazier samples and rock core from the site investigation.

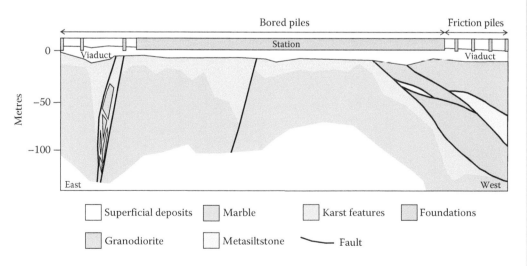

Figure CS12.1.3 Simplified geological cross-section through Yuen Long railway station. (After Arup, 1998.)

Figure CS12.1.4 Site investigation samples from vertical boreholes, Yuen Long Station. (a–d) SPT liner samples. (e) Mazier sample and (f) rock core. (a) Marine mud and estuarine silty sand. (b) Alluvium, silty sand and fine sand. (c) Supra-karst deposits against jointed marble, sub-vertical contact. (d) Finely bedded cavity infill silt and sandy silt. (e) Completely decomposed granodiorite overlain by alluvial sand. (f) Massive marble with dissolution features along sub-horizontal joints.

Bored piles end-bearing piles were used for both the station and eastern viaduct foundations, whereas bored friction piles were employed at the western end in the completely decomposed granodiorite.

ENGINEERING CONSIDERATIONS

- Marble sequence is displaced by thrust and normal faults.
- A tabular body of granodiorite has been thrust over the marble sequence.
- Karst formation is most intense close to the faults.
- Filled and possible open cavities are present within the marble.
- Weathering, including karst features and decomposition of the granodiorite, extends to over −120 mPD.

CASE STUDY 12.2   KARST AND DEEP WEATHERING: APARTMENT BLOCK, TUNG CHUNG, HONG KONG

[22°17.57′N 113°57.07′E]

A new town was constructed on a reclamation at Tung Chung, Lantau Island, close to the Hong Kong International Airport. The proposed 50-floor, Apartment Block 5 was never built (Figure CS12.2.1), due to the steep inclination of rockhead across the building footprint, the presence of loose cavity-fill materials and possible open cavities at extreme depths (Figure CS12.2.2). These complex geological conditions (Fletcher et al., 2000; Kirk, 2000; Wightman et al., 2001) were largely unsuspected when the reclamation was first planned, as the mapped geology of the adjacent onshore areas is composed of mainly granites and volcanic rocks (Figure CS12.2.3).

Figure CS12.2.1 Apartment blocks during construction at Tung Chung. The gap marks the planned position of Apartment Block 5.

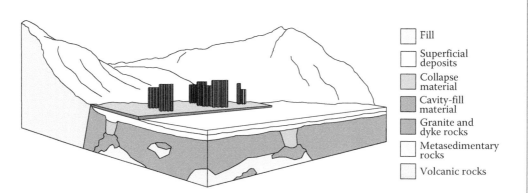

Fill

Superficial deposits

Collapse material

Cavity-fill material

Granite and dyke rocks

Metasedimentary rocks

Volcanic rocks

Figure CS12.2.2 Schematic block diagram of the geology of the Tung Chung area.

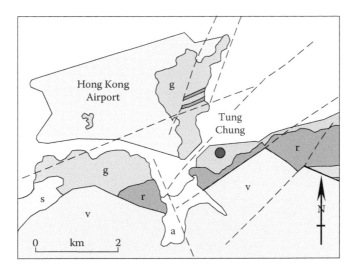

Figure CS12.2.3 Geology of Tung Chung area prior to site investigations. v, volcanic rocks; g, granite; r, rhyolite dykes; s, sedimentary rocks; a, alluvium; grey, reclamation. Dashed lines – interpreted faults. (After Geotechnical Engineering Office, *Geological Map of Hong Kong*, Millennium Edition, 1:200,000 scale, Civil Engineering Department, Hong Kong SAR Government, 2000.)

GEOLOGY

The mountains behind Tung Chung are composed mainly of volcanic rocks (Figure CS12.2.3) which are in faulted contact with complex ground conditions that consists of metasedimentary rocks, rhyolite dykes, granite, cavity-fill sediments and other karst-related deposits. These materials are overlain by marine clays and alluvium. Tropical weathering over tens of thousands of years, when sea levels were much lower than today, has extended to over 150 m below present sea level. Rockhead is depressed from an average of about −60 mPD across most of the reclamation to over −150 mPD beneath Apartment Block 5 (Figure CS12.2.4). Continuous sampling of the soils within this deeply weathered zone at depths between −100 and −150 mPD recovered laminated clay, silt, sand and sedimentary breccias (Figure CS12.2.5a and b). The soils were recovered using polymer drilling fluids as conventional methods did not allow recovery of such weak and friable material. The presence of these variable soils, the steep inclination of the soil/rock interface and the risk of encountering cavities within the marbles at this site resulted in the abandonment of the construction of this apartment block (Figure CS12.2.1). The possibility of using bored piles was contemplated, but the depth to bedrock was considered impractical and too expensive. The adjacent blocks were either founded using bored piles into bedrock or driven piles into homogeneous soils (Figure CS12.2.6).

These sediments have been interpreted to have been deposited within a cavity that formed during the last glacial period in a large marble block (Fletcher et al., 2000; Wightman et al., 2001). The former cavity was at least 30 m across and 50 m high and is probably completely filled with sediment. The floor of the cavity is composed of slightly decomposed and fresh metasedimentary rock, skarn, rhyolite and hydrothermally altered granite. Large, completely decomposed, granite blocks set in a variable sedimentary breccia, clay and sand matrix overlie

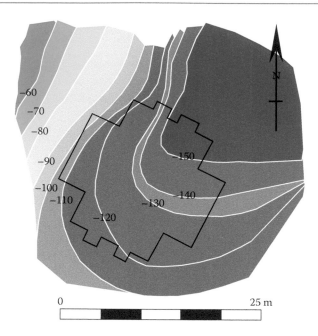

Figure CS12.2.4 Rockhead contours beneath the footprint of Tower Block 5 (outlined in black). Depths in metres.

Figure CS12.2.5 Mazier samples of cavity-fill material. (a) Thinly laminated clay and silt with a few sand layers – lacustrine sediments. (b) Sedimentary breccia – roof-collapse deposits.

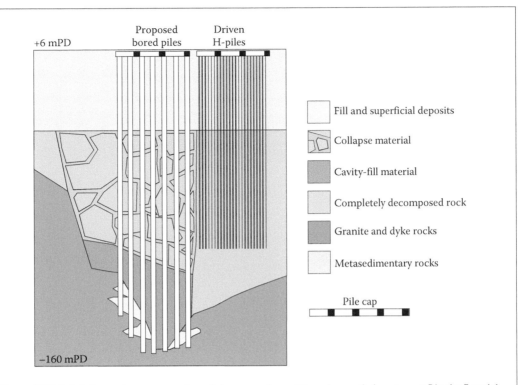

+6 mPD

Proposed
bored piles

Driven
H-piles

−160 mPD

Fill and superficial deposits

Collapse material

Cavity-fill material

Completely decomposed rock

Granite and dyke rocks

Metasedimentary rocks

Pile cap

Figure CS12.2.6 Schematic cross-section of the ground conditions beneath Apartment Blocks 5 and 6.

the cavity-fill sediments. These deposits were formed during the progressive decomposition and collapse of the granite roof to the former cavity.

ENGINEERING CONSIDERATIONS

- Cavity-fill material encountered at considerable depth.
- Steeply inclined rockhead profile that is depressed to over −150 mPD in places.
- Loose medium sands within the former cavity easily disturbed during piling.
- Abrupt transition from dense decomposed granite to loose sand.
- Fundamental change in bedrock geology from onshore to offshore.

Chapter 13

Superficial deposits

Superficial deposits are mainly unconsolidated sediments that have accumulated on bedrock, weathered bedrock or earlier superficial deposits during the last tens of thousands of years and in places are still being deposited today. Climate warming since the end of the last ice age has greatly influenced the form and composition of superficial deposits because of the migration of depositional environments and the quantity of transported sediments over time. For example, the retreating glaciers in the Himalaya leave moraines in the sculptured valleys (Figure 13.1) that are then eroded, transported and deposited as alluvium in the lower reaches of the same valley that was once filled by a glacier. Sequences through many superficial deposits provide a detailed record of changing environments resulting from climate warming, tectonic uplift and/or sea level variations. Analysis of these sequences allows robust geological modelling of the ground conditions for engineering works and can establish the architecture of the deposits in relation to geotechnical parameters, hydrogeology and locations of potential risk.

Superficial deposits can broadly be divided into two groups: those that accumulated in a continental environment and those of oceanic and nearshore environments (see Chapter 4). The first group includes sediments deposited in river systems (*alluvium*), on hillslopes (*colluvium*) and in lakes (*lacustrine deposits*), together with those associated with the action of wind (*aeolian deposits*) and glacial events (*glacial deposits*). The second group includes deep oceanic mud, deltaic deposits, estuarine mud and silt and shoreline and shallow-water deposits. At any one time within a particular area, a range of superficial deposits can accumulate in different geomorphological environments. For example, Figure 13.2 presents a schematic cross-section of the geometric relationships between onshore and offshore superficial deposits in Hong Kong since the last glacial period. Old colluvium, deposited on hillsides during and shortly after the last ice age when sea levels were low, graded at lower elevations into thick successions of alluvium that spread across the valley floors. In postglacial times, as sea levels rose, the coastal areas were submerged, and the earlier terrestrial deposits were covered by marine sand and mud, whereas on higher ground, colluvium continued to accumulate.

Recent superficial deposits have their counterparts in the geological record and therefore share many of their features. For example, recent deltaic sediments have the same form, sedimentary structures and grain composition as those encountered in ancient deltaic sedimentary rocks. However, in sedimentary rocks the grains are cemented by variable quantities of minerals such as quartz and calcite, whereas in recent deposits the cementation is much less intense or absent. As a consequence, descriptions of recent oceanic and deltaic sediments are not included in this chapter, as their characteristics are described in Chapter 4.

Figure 13.1 End moraine across a glaciated valley to the north of Mount Qomolangma (Everest), Tibet. Note that much of the finer-grained material is in the process of being eroded and transported from the side of the deposit.

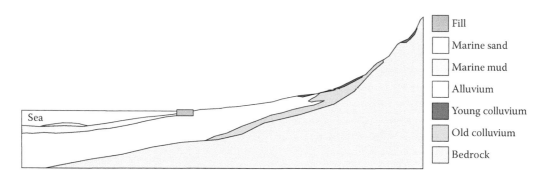

Fill
Marine sand
Marine mud
Alluvium
Young colluvium
Old colluvium
Bedrock

Sea

Figure 13.2 Schematic cross-section illustrating the geological setting of onshore and offshore superficial deposits. (After Fyfe, J.A. et al., 2000, Published with permission of the Geotechnical Engineering Office, Civil Engineering Department, The Government of the Hong Kong SAR.)

ENGINEERING CONSIDERATIONS

Superficial deposits in engineering terms are generally unconsolidated soils with little or no cohesion, and some may act as aquifers or aquicludes, thereby greatly affecting the design and construction of tunnels, foundations, dams and other engineering works. The heterogeneity of superficial deposits makes their compressibility highly variable and their hydrogeological flow patterns complex and changeable. Many superficial deposits are being deposited in active dynamic sedimentary systems today, and knowledge of how their architecture has changed over time must be fully appreciated. This change may be in the form, for instance of the encroachment of windblown sand, failure and erosion of hillslope deposits and channel migration in river systems. In addition, present-day or recent surface and near-surface processes may radically alter the geotechnical properties and responses of these deposits, for example precipitation of iron and calcium mineral cements and concretions, desiccation, development of sulphate segregations, formation of soil pipes and the leaching of soluble components (e.g. Case Study 11.2). Climate also has a large influence on how the properties of superficial deposits change over time, ranging from intense biodegradation in wet tropical climates to freeze–thaw disturbance in arctic and subarctic environments.

Sampling of superficial deposits during site investigation for ground engineering projects is generally undertaken using Vibrocore, Mazier and standard penetration test (SPT) liners, which can provide relatively undisturbed complete sections through the deposits. However some samples may show minor disturbance along the margins of the sample due to friction of the material against the plastic or metal sampling tubes, for example bedding in the alluvium sample shown in Figure 13.4c has been bent downwards by the sampling procedure. It is essential that as many as possible of these samples, except for those needed for geotechnical testing, are opened to establish the composition of the sediments, bedding characteristics, structure, nature of any secondary cements and hydrogeological information, including the potential for changes in permeability and the presence of soil pipes. Examples of superficial deposits collected during site investigations in Hong Kong are included in this chapter to highlight important characteristics of each of the samples that could have been missed during normal logging procedures. To obtain the maximum recovery of very soft superficial deposits during the sampling process, the use of specialized drilling fluids, such as polymers, has been found to be beneficial.

It is impossible to summarize all the engineering considerations of the different types of superficial deposits here, as they are so variable and dependent on local depositional conditions. However, some of the main geological factors that should be considered during the formulation of the geological and ground models are as follows:

Alluvium

- Interlayers of gravel, sand and silt with fan, channel, sheet and lens forms.
- Thickness variations are dependent on the old topography.
- Hard desiccated crusts form during exposure to the atmosphere, and these can influence water flow patterns within the alluvial sequence.
- Alluvium is deposited in a very dynamic environment with frequent changes in location and morphology of the component sediment bodies.

Colluvium

- Composed of a poorly sorted mixture of soil and rock fragments.
- A high percentage of boulders in colluvium may increase the mass shear strength.
- Recent colluvium is often highly permeable and may contain soil pipes.
- Older colluvium may be compact and contain secondary clay minerals such as kaolinite.

Nearshore marine deposits

- Composed largely of very soft to soft clayey silt and mud with, in places, a basal sand layer that grades into coastal shoreline deposits.
- Largely accumulated during the last rise in global sea level.
- Highly compressible in part and this can result in large settlements.
- Marine mud layers can act as impermeable barriers to water flow.

Glacial deposits

- Many glacial deposits consist of unsorted, angular fragments of rock set in a silty clay matrix.
- Moraines are generally very loose and contain a wide variety of clast sizes.

- Large isolated boulders may occur within finer-grained material.
- Glacial lake deposits are generally soft and clay-rich with thin interlayers of sand and gravel.

Engineering projects in glaciated terrains can be extremely complex due to the varied nature and distribution of the glacial deposits, which can range from mounds of boulders, lenses of sand and gravel, to clay-rich layers with isolated large boulders. Variability of the geotechnical properties of these glacial deposits combined with complex hydrogeological regimes provides challenges to construction and tunnelling projects. In addition, the frozen ground in permafrost regions can create severe problems for foundation stabilization, especially as the global temperatures continue to rise.

Three case studies of ground engineering projects are presented in this chapter, which illustrate the complexity of superficial deposits and the necessity for a full understanding of the processes that formed them. The first case study presents the results of the ground investigations for bridge foundations in Bangladesh across the Padma River, which passes upstream into the Ganges drainage system (Case Study 13.1). Here, the composition and densities of the different layers in the thick alluvial sequence reflect the variable erosion rates of the Himalaya mountains during global climate change since the last interglacial period. The second case study describes a landslide in Wales where a sequence of thinly bedded glacial lake clays failed as a result of recent changes in slope morphology (Case Study 13.2). The third case study relates to a preliminary investigation into the feasibility of using loess (windblown dust) for earthworks in the large expansion of an energy development in Kazakhstan (Case Study 13.3). The reader is also referred to Case Study 2.1, which details the composition and architecture of the glacial and postglacial superficial deposits beneath a reclamation for Hong Kong Airport, and Case Study 2.4 which describes a contact between colluvium and weathered volcanic rocks.

ALLUVIUM

Alluvium is the general term used to describe all the various sediment types that were deposited in river systems. The component material is predominantly transported in river water by traction currents that carried the dispersed grains in suspension or moved them by rolling along the river bed (Miall, 1992). More rarely alluvium may form as debris flows, which are chaotic masses of saturated sediment that move rapidly down the river channel under extreme rainfall conditions.

The variation in the features of alluvial deposits is determined by the geomorphology of the river system in which they were deposited, which in turn is controlled by the overall gradient of the river. This may vary from narrow dendritic channels filled with coarse gravels and boulder beds in the upper parts of a river system to alluvial planes with meandering channels of sand flanked by swathes of silt and mud. This transition is well displayed in the satellite image from the eastern sector of the Himalaya (Figure 4.4a) and is described more fully in the 'Continental environment' section of Chapter 2. In places, *levees* or mounds of coarser material may build up along the riverbanks, which are only breached at times of flooding, for example along the Mississippi River, near New Orleans, United States. Abrupt changes in river gradient and/or widening of the drainage pathways will lessen the velocity of the water flow and will result in the lowering of sediment load parameters and the consequential deposition of the suspended material. For example, a cone of sediment will form at the mouths of narrow river valleys as they open out onto flatter ground; this is called an *alluvial fan*.

(a)

(b)

(c)

(d)

Figure 13.3 Alluvial environment. (a) Incised steep-sided valley with alluvial deposits of gravel with boulders, eastern Nepal. (b) Thick alluvial sand and gravel exposed in a series of alluvial terraces related to the downcutting of an incised river meander, Nepal. (c) Braided river valley with deposits of light grey, volcanic silt and ash, Mount Pinatubo, Philippines. (d) Alluvial fan composed of a sequence of debris flow deposits and water transported sand and gravel, Himalaya, central Tibet.

Examples of alluvial environments include narrow mountain valleys (Figure 13.3a), incised *alluvial terraces* that reflect changing deposition/erosion events (Figure 13.3b), migrating braided streams across a wide alluvial valley (Figure 13.3c) and an alluvial fan composed of predominantly of debris flow deposits (Figure 13.3d). The composition of alluvial sediments is highly variable, both laterally and vertically, and largely dependent on river morphology, river gradients and water flow rates. However, seasonal or long-term global climate changes will strongly influence alluvium characteristics, in that the water flow rates in the drainage systems are largely determined by variations in rainfall and/or the melting rates of snowfields and glaciers. Thus, alluvial sedimentary sequences can provide a detailed chronology of past climates. For example, recent alluvial sequences on the flanks of Mount Pinatubo volcano, Philippines, are composed of well-defined beds of rounded boulders, fine sand and gravel, each reflecting sudden changes in rainfall intensity (Figure 13.4a). The geotechnical parameters of the individual alluvial units can be markedly different, and therefore their architecture and composition must be thoroughly understood in any engineering project. This is illustrated in Case Study 13.1 beneath the Padma River flood plain, where some 2000 m of alluvium has been deposited during glacial and interglacial periods.

(a)

(b)   (c)   (d)

Figure 13.4 Alluvium. (a) Riverbank exposure of alluvium composed of boulder-rich beds interlayered with beds of pebbly sand, silt and mud, Mount Pinatubo, Philippines. (b) Bedded, light brown, sandy silt and coarse sand. SPT liner sample, Hong Kong. (c) Finely bedded silt and sandy silt. Mazier sample, Hong Kong. (d) Mottled, iron-stained desiccated crust passing downwards into light-grey coarse sand. SPT liner sample, Hong Kong.

The composition, grain size and morphology of alluvial bodies are highly variable, and therefore, complete sampling of these deposits, as far as possible, during any site investigation is paramount, as they will also provide additional information on the stiffness, permeability and compressibility of the sediments. Some examples of alluvium from SPT liner, Vibrocore and Mazier samples from Hong Kong and Bangladesh are illustrated in Figures 13.4b–d and CS13.1.5.

COLLUVIUM

Colluvium is a general term for loose heterogeneous deposits of soil and rock fragments that accumulate by mass-wasting processes on or close to the base of steep slopes or cliffs. Colluvium includes landslide deposits, hill-creep accumulations and talus, and can imperceptibly grade or be interbedded with alluvium on the valley floor. Most colluvium deposits are restricted to the land, but in places they may extend a short distance into marine and lacustrine environments or be overlain by later subaqueous deposits. On upper slopes, colluvium rarely displays any internal bedding features or sorting of the rock fragments, but at lower elevations, some sorting of the rock fragments can take place, and isolated sand and silt lenses may constitute a poorly defined bedding. Colluvium may be composed of layers of material that were deposited at different times and under a range of environmental conditions. For example, a layered sequence of colluvium in Hong Kong was deposited on weathered volcanic tuff during distinct depositional events (Figure CS2.4.1).

Some of the common depositional environments of colluvium are illustrated in Figure 13.5. *Talus* or scree consists of a chaotic assortment of fallen, angular rock debris at and close

(a)

(b)

(c)

(d)

Figure 13.5 Colluvium environment. (a) Talus slope composed of angular blocks that have fallen down from a cliff of quartzite, Holyhead, Wales. (b) Fine-grained slope deposits below cliffs of bedded tuff and siltstone that have been incised by a recent drainage system. A slight layering of the colluvium is exposed in a small cut at the bottom left of the photograph, central Tibet. (c) Angular debris derived from a large landslide, Hope, Canada. (d) Thin layer of colluvium resting on weathered volcanic rocks, exposed in the back scarp of a small landslide. The colluvium is formed of fractured bedrock that accumulated mainly by hill creep, Tai O, Hong Kong.

to the base of cliff faces or on very steep rocky slopes (Figure 13.5a). In cold climates, the fracturing of the rock is largely controlled by freeze–thaw action combined with dislodgement of the rock fragments by rain. The rock debris is very loose with the angles of repose ranging from 40° to 60°, which is dependent on the size and shape of the rock fragments. Some poor sorting of the rock fragments may occur down the length of the scree. Where the cliffs are more friable and weathered, the colluvium is finer grained and spreads as an apron below the cliffs. Slope movements within such a soil mass and gravity grading of the rock fragments may result in the development of an ill-defined layering in the lower parts of the colluvial deposit (Figure 13.5b).

Landslides produce large volumes of colluvial debris that come to rest either on the hill-side or spread out across the lower ground. Landslides are a major factor in the erosion and geomorphological development of open hillsides, for example the immense fatal landslide near Hope, British Columbia, Canada, in 1965 deposited a 70 m thick mound of debris in the valley (Figure 13.5c).

Fragmentation of bedrock and slow creep down steep hillsides can form thin blankets of colluvium across open hillslopes (Figure 13.5d). On moderate slopes, there is a tendency for the surface material to move slowly downhill under gravity due to sliding between constituent grains without the complete loss of cohesion. Under steeper conditions or water pressures, cohesion is lost and failure occurs. In places, the creep deforms the discontinuities in the underlying rock mass, for example the downhill bending of cleavage planes in a slate beneath a layer of hill-creep material. At times of high rainfall, the surface layers become more mobile and form what is termed *hill-wash*. Vegetation helps to stabilize slopes susceptible to hill creep, and recognition of deformed tree trunks may point to the influence of hill creep over the time of the tree's growth.

Recent colluvium is generally highly permeable and consists of angular cobbles and gravel set in finer grained sandy silt matrix. Post-depositional cementation and mineralogical changes, particularly oxidation of the clay/silt and decomposition of the rock fragments, are ongoing due to groundwater flow through the body of the deposit or in soil pipes and exposure of the surface of the colluvium to the atmosphere. Where the open spaces are present between the rock fragments, some redistribution of the fine materials by groundwater may take place. Over time, colluvium becomes more indurated, and in tropical climates where the oxidation and rock decomposition are more intense, clays such as kaolin may form (Figure 13.6b). Colluvium may accumulate on fresh and slightly decomposed rock (Figure 13.6a), on completely decomposed soils at the top of a weathering profile (Figure CS2.4.1) and on older colluvial deposits.

There is great difficulty in obtaining good continuous samples of colluvium by normal site investigation drilling procedures, as the material generally consists of a heterogeneous mixture of rock fragments of variable size and soil. However, where sections of the colluvium contain small rock fragments, it is possible to recover sections of soil in SPT liner and Mazier samples (Figure 13.6c and d).

Erosive action of water flowing through colluvium or completely decomposed rock may form soil pipes that range in size from a few centimetres to over a metre. The soil pipes may follow relict joints in completely decomposed rock and can be present along the rock–soil interface. In colluvium, they are less systematic and find pathways around the numerous rock boulders. Although open pipes provide pathways for water into the body of the superficial deposits or decomposed rock, many are filled with sand and silt washed in from the ground surface. Such material is regarded as a superficial deposit although found within the body of the soil.

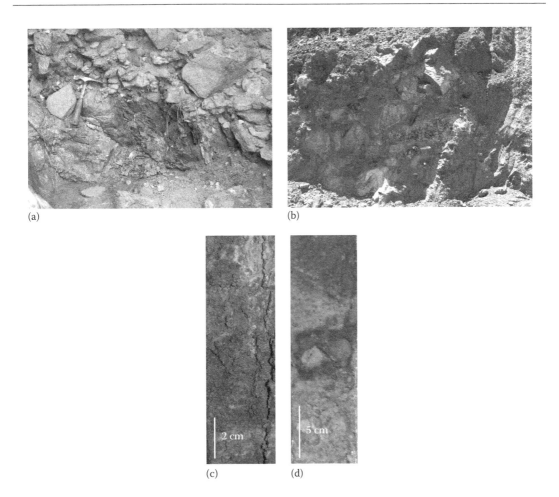

Figure 13.6 Colluvium from Hong Kong. (a) Unsorted, angular cobbles of local bedrock in clayey sand overlying moderately weathered, foliated, volcanic tuff. (b) Colluvium with large angular blocks of completely decomposed granite in a silty sand matrix. Decomposition of the granite blocks and induration of the matrix suggest that the colluvium is relatively old; it is overlain by a more recent, loose colluvium seen in the upper left of the plate. (c) Colluvium composed of brown, very silty sand with rare rock pebbles. SPT liner sample. (d) Slightly bedded colluvium with reddish sandy layer overlain by pebble-rich, sandy clay.

SHALLOW-WATER MARINE DEPOSITS

Marine deposits that are discussed here are those that have been laid down on the seafloor close to the shoreline and exclude those of the deeper offshore regions. The nearshore environment has been highly changeable and largely controlled by a rise in sea level following the last ice age, tidal and current influences and the frequency of storm surges. In places, the complex distribution of offshore superficial deposits over relatively small areas makes ground modelling challenging. Such variations are particularly important in the formation of land reclamations, where differential compaction may take place and hydrogeological flow patterns are changeable. The shallow-water marine deposits of Hong Kong have been studied in great detail over the last few decades as the number of reclamations has increased

(a)   (b)

Figure 13.7 Marine deposits from Hong Kong. (a) Grey, marine mud with few shells, weakly laminated. Vibrocore
    sample. (b) Grey, marine mud overlying mottled alluvium composed of medium sand. Mazier sample.

to provide sites for new towns, harbour facilities and an airport. The distribution of the
nearshore marine deposits has relied mainly on seismic reflection surveys (Fyfe et al., 2000).
This is exemplified by the detailed geophysical surveys which were undertaken prior to the
establishment of the reclamation platform for the Hong Kong Airport (Plant et al., 1998).
Case Study 2.1 presents the main features of the marine and other superficial deposits at this
site. A thorough knowledge of the composition and distribution of marine deposits has also
been paramount in the construction of immersed tube tunnels, waste disposal sites and the
location and extraction of marine sand for reclamations.

Marine deposits can readily be recovered in Vibrocore, Mazier and SPT liner samples,
which provide details of sedimentary structures and bedding features. For example, fine
laminations in marine mud can be preserved (Figure 13.7a) and the nature and exact location of the boundary between the marine mud and the underlying superficial deposits
defined (Figure 13.7b).

WINDBLOWN DEPOSITS

Windblown deposits include localized coastal and inland dunes, extensive desert sands and
loess (Brookfield, 1992). They have formed by the action of wind carrying sand grains or

Figure 13.8 Aeolian deposits. Sand dunes within a wide alluvial valley, central Tibet. The sand is blown into the lake and there becomes part of the lacustrine sediments.

clay-sized particles by strong winds close to the ground or in dense clouds that may reach heights of over 10,000 m. The sand grains in the windblown deposits are almost completely spherical due to abrasion of the grains against one another. In rocks, such spherical grains provide conclusive evidence for the depositional environment of certain sandstones (Figure 4.5a). Windblown sands may form thin beds in lacustrine successions where the sand dunes are proximal to the lake (Figure 13.8).

Coastal and inland dunes

Coastal dunes commonly form parallel to the shoreline just above the high tide level. Prevailing winds dry out the beach sands at low tide and carry the sand grains inland, in some places for several kilometres, where they accumulate as sand hills. Some sand hills are relatively static and reach heights of over 100 m, whereas others are continually changing shape and on the move. Commonly, the encroachment of coastal and inland dunes occurs at times when the climate was favourable for their formation, perhaps times of low rainfall or during cold winters. For example, along the south coast of Wales, near Porthcawl, sand dunes covered a Roman village during the fifteenth and sixteenth centuries, but since that time have not moved. In arid mountainous regions, isolated accumulations of windblown sand can form thick deposits on exposed hillsides. In Tibet, close to Lhasa, the east-facing slopes are covered with blankets of sand driven by the prevailing easterly winds, and small sand dunes have been established on a wide alluvial plane (Figure 13.8). In both cases, the amount of available sand and the rocky terrain have hindered the development of extensive sand accumulations.

Desert sands

Only about 20% of the world's deserts are covered by windblown sand, and of these, most are concentrated in the North African and Middle Eastern regions. There, the sand is piled up into a plethora of dunes of varying morphologies, including crescent-shaped barchans, linear dunes and star-shaped dunes, which are dependent on the strengths and directions of the prevailing winds. In the context of this book, the most important aspect of progressive desertification of a region is the encroachment of sand across fertile land and into urban areas, as well as interfering with transport networks and other infrastructure projects. Desertification results mainly from deforestation and poor farming practices, but other factors such as lack of rain caused in part by climate change almost certainly play a significant role. Projects in these areas may have to consider ways in which to ameliorate the advance of the deserts by the establishment of vegetation barriers.

Loess

Loess is windblown, porous material that has accumulated as thick blankets across parts of central Europe, North and South America and Asia. The most extensive deposits of loess occur in central China where they have attained thicknesses of over 100 m during the Quaternary Period, particularly in the interglacial intervals. It is generally buff coloured and very homogeneous with no or little stratification, but may include thin, laminated sections (Figure CS13.3.3). Loess is composed mainly of calcareous silt and clay and has a weak cohesion that enables the loess to be dissected by steep-sided gullies. Close to ground surface calcareous nodules may develop in the loess as calcium-bearing groundwaters percolate through the soil (Figure CS13.3.4).

An engineering project in a loess related to the construction of an energy development in Kazakhstan describes the composition, mineralogy and geotechnical properties of loess (Case Study 13.3).

LAKE DEPOSITS

Lake, or lacustrine, deposits are for the most part very similar to the sediments that accumulate in shallow-water marine sequences, in that both include shoreline and deltaic environments, although in lakes the extent of these environments tends to be much more restricted. Also in lakes there is an absence of currents, tides and carbonate accumulations so that the sedimentary structures and deposits related to these are absent. Lakes on the other hand provide a relatively calm shallow water and closed depositional system, which can preserve minor changes in sediment input as revealed in minutely layered successions (e.g. Figure CS13.2.5b). The changes may be controlled by rainfall patterns, temperature variations, evaporation or even biological factors and therefore provide a detailed environmental history of the whole region. Lake deposits have been exposed around the shores of a lake in Tibet (Figure 13.9a) as the lake levels were lowered, possibly due to evaporation, reduced

(a)   (b)

Figure 13.9 Lake deposits in Tibet. (a) An exposed sequence of thinly bedded clay and silt that were deposited in a lake when water levels were much higher. (b) Evaporite deposits forming within a high-altitude lake due to low annual rain and snow fall, dry winds and evaporation.

rainfall and/or tectonic uplift. The layered deposits of mud, silt and sand are exposed in the banks of incised streams that drain into the lake.

EVAPORITE DEPOSITS

Evaporite deposits are chemical sediments that form by precipitation of brines from seawater or lake waters in the continental setting (Kendall, 1992). The most common evaporite minerals are gypsum ($CaSO_4.2H_2O$) and halite (NaCl), but there are a host of other Ca, Na, K and Mg minerals. Evaporite deposits form within isolated seas, lakes and lagoons where precipitation of the minerals is controlled by the increased concentration of the ions within the brine. Thick beds of evaporite within a geological sequence can migrate through the overlying strata to form salt domes. Evaporite deposits have been mined extensively, and this can lead to collapse features at surface.

An example of salt precipitation around the shores of a shallow lake in Tibet is shown in Figure 13.9b. As water influx rates are low, this lake will eventually be completely filled with evaporite deposits, as has happened with the much larger lakes of the Bonneville Salt Flats, Utah, United Sates [40°33′N 113°35′W] and the Salar de Uyuni, Potosi, Bolivia [20°6′S 67°32′W].

GLACIAL DEPOSITS

The Pleistocene epoch, which lasted from approximately 2.6 million to 12,000 years ago, was a time of repeated cold periods when the polar ice, mountain glaciers and ice sheets were at their maximum coverage and thickness. In places, the ice sheets were 2000 m thick, and in the northern hemisphere, they reached as far south as the Great Lakes in North America and London in Europe. There were four main glacial periods, and during those times, ice covered nearly 30% of the Earth's surface and sea levels were lowered by up to 120 m. As the ice retreated during the interglacial periods and since the last ice age, a large variety of glacial debris was deposited in glaciated valleys and in front of the ice sheets. The glacial deposits and their topographic signatures are preserved in areas once covered by ice sheets and valley glaciers, for example in Scotland and Canada.

An excellent way to understand glacial processes and landforms is to view and explore for oneself satellite images of glaciers (Figure 13.10) using Google Earth and reference these with standard texts (e.g. Summerfield, 1991). Two images are presented from the east coast of Greenland where glaciers sourced from the central ice cap flow down valleys into *sea fjords* (drowned valleys): The first satellite image (Figure 13.10a) shows many glacial features, including dark bands of entrained rock fragments, which define lateral and medial *moraines*, flow fold deformation of the moraine trails close to the glacier front and floating ice debris in front of the melting glacier. The second satellite image (Figure 13.10b) shows retreating glaciers that have left *U-shaped valleys* partially covered by moraines, colluvium and alluvium. An ice-dammed lake has formed in one of the side valleys, and this will be gradually filled with fine sediment during the summer months when not covered by ice. A small glacier has flowed into the main valley depositing glacial debris onto the alluvial plane. These local depositional environments result in highly complex architectures to the superficial deposits. Where the glaciers calve into the sea it is sometimes possible to see stratified, entrained rock debris within the margins and bases of the glaciers (Figure 13.11a).

Many glacial sequences are composed of unstratified, silty clay containing numerous unsorted, angular boulders and cobbles (Figure 13.13a). A general term for these deposits assumed to be laid down by ice is *till* or *boulder clay*. However, it is now considered more

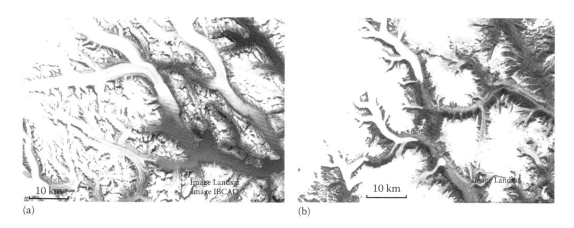

(a)   (b)

Figure 13.10 Glaciers descending from the Greenland ice cap. (a) Valley glaciers breaking up as they enter sea fjords. Lateral and medial moraines are defined by dark boulder trails within the glaciers. Google Earth, Landsat image [63°06'N 42°22'W]. (b) Retreating glaciers leaving U-shaped valleys partially covered by moraines, colluvium and alluvial deposits. A small, ice-dammed lake can be identified in the lower central part of the image. Google Earth, Landsat image [74°39'N 22°13'W].

(a)   (b)

Figure 13.11 Glacial environment in Alaska, United States. (a) Layered rock debris at the margins and along the base of the Margerie Glacier. (b) Seasonal lake at the front of the Mendenhall Glacier.

appropriate to use the completely non-generic term *diamict* to describe these deposits as they might form under markedly different conditions. Glacial deposits on land are grouped into four categories: subglacial material close to the base of the glacier (lodgement tills, drumlin diamict, esker sand and gravel), supraglacial material originally on top of the glacier (moraine diamicts), glaciofluvial outwash sediments in front of the glacier (gravel and sand) and glacio-lacustrine sediments in lakes (varved clays, delta sands) (Eyles and Eyles, 1992).

Following the retreat of the glaciers and shrinking of the polar ice caps, the temperatures remained very low across the high latitudes and in elevated regions. At shallow depths, the ground remains permanently frozen (*permafrost*), but freeze–thaw conditions close to exposed surface initiate heave, cracking and mass movement of the ground. Engineering

(a)   (b)

Figure 13.12 Glacial features. (a) Lateral moraines deposited along the valley sides, at the Everest base camp, Tibet. (b) Drumlins and lateral moraines left by a retreating glacier, Southern Tibet.

projects in these regions require special considerations, especially as the depth to the permafrost is increasing due to global warming.

Moraines

Moraines are unconsolidated glacial deposits that form subglacially, supraglacially or at the ice margins (Summerfield, 1991). They commonly form distinctive topographic features such as elongate mounds and for the most part consist of a mixture of unsorted angular boulders set in a sandy and silty sand matrix, best referred to as diamict (see earlier). Moraines are commonly reworked by meltwater processes and general slope degradation on the sides of the deposit (Figure 13.1).

Glaciers pluck boulders and scour finer material from the floors of glaciated valleys, and rock fragments persistently fall from the valley walls onto the glacier. The entrained boulders form dark bands along the edges of many glaciers (Figure 13.11a) (*lateral moraines*), and where two glaciers merge, their lateral moraines coalesce to form a band in the middle of the resultant larger glacier (*medial moraine*). These moraines are brilliantly displayed on the satellite images of west Greenland (Figure 13.10a). Following the retreat of a glacier, the mounds of moraine are deposited in front of glacier (*end moraines*) or along its flanks. To the north of the Mount Qomolangma (Everest), Tibet end moraines form ridges across the valley floors (Figure 13.1) that consist of a chaotic mixture of angular blocks of rock, gravel, clay and sand scoured from higher parts of the mountain. Lateral moraines are also prominent in most of the valleys of the region (Figure 13.12a).

Meltwater deposits

Meltwaters that flow out from beneath glaciers commonly form outwash fans in front of the retreating glacier, either into a proglacial lake or across the valley floor. Their depositional characteristics vary from debris flow to water-lain deposits, the former having little or no internal structure, the latter showing some evidence of sedimentary layering (Figure 13.13b). Outwash gravels (Figure CS13.2.5a) were identified in the site investigation of a landslide in Wales (Case Study 13.2), and a schematic diagram of their environment of deposition in relation to the glacier front (Figure CS13.2.4) was constructed (Fletcher and Siddle, 1998). Alluvial deposits derived from reworked, glacial debris stretch down the

(a)   (b)

(c)   (d)

Figure 13.13 Glacial deposits of the last ice age, Wales. (a) Glacial diamict composed of angular cobbles and gravel set in a silty clay matrix, St Dogmaels. Width of trial pit sample 15 cm. (b) Water-lain, glacial sand with pebbles deposited on a glaciated rock surface, Traeth Llydan. (c) Slightly layered, boulder-rich silty sand within a drumlin, Hen Borth. (d) Large, erratic boulder of igneous rock (gabbro) within glacial deposits. The original surfaces of the boulder, where not broken off, are smooth and facetted due to abrasion at the base of a glacier, Porth Nobla.

valley in front of the retreating glacier (Figure 13.10b), their architecture being determined by braided and anastomosing river systems that change their morphology throughout the year. Sand and gravel accumulates in transient channels or spreads as sheets across the valley floor, dependent on the season. Due to the lack of vegetation, windblown sands may also accumulate on the alluvial plain and swamp conditions may develop abandoned river channels.

Glacial lake deposits

In front of glaciers, it is common for lakes to form that are fed by the meltwaters from the glaciers (Figure 13.11b). In the summer months when melting of the glacier is at its height, the sediment entering the lake is fine sand and silt, whereas in the winter months, the lakes are frozen over and the only sediment being deposited is the suspended clay from the water column. These annual sediment couplets of silt and clay are called *varves* (Figure CS13.2.5b and c) and accumulate over time into thick, laminated deposits. The variation in thicknesses of the sand and clay couplets tracks the changes in climate over thousands of years.

Drumlins and eskers

A feature of many previously glaciated terrains is the presence of numerous hummocks of diamict (*drumlins*) and steep-sided, meandering ridges (*eskers*). Both were originally formed beneath the glacier: drumlins as deformed and moulded till beneath the glacier and eskers as sediment infills of subglacial streams. Drumlin mounds have lens shapes (Figure 13.12b), which are orientated parallel to the ice movement. They are composed of diamict with whisps and lenses of more sandy material, which indicates some reworking of the till by subglacial waters (Figure 13.13c).

Erratic boulders

Large, isolated boulders are found within glacial or subaqueous deposits. These may have been either deposited directly by glaciers or carried by floating ice, in the form of icebergs or ice sheets, great distances from the glacier that once carried them. These *erratic boulders* are distinctive in that they are very large in comparison with the components of the enclosing materials in which they are found, have smooth, facetted edges and commonly are sourced from outside the immediate area. A single, large erratic boulder exposed in a low, coastal cliff of glacial diamict on Anglesey, Wales, displays these features (Figure 13.13d).

PEAT

Peat is the accumulation of decayed vegetation over upland moors, in forests and in wetlands since the retreat of the polar ice at the end of the last ice age. It is composed of a wide variety of plant materials, commonly sphagnum moss, but can also contain leaf compost, twigs and pine cones that formed the ground cover to the woodlands. Generally, exposed peat is about 2 m thick and has been cut for fuel over the years and is now used as garden compost. Around the coast of Wales, a peat layer is occasionally exposed at low tide across beaches (Figure 2.7a and b). It consists of tree stumps and roots embedded in organic matter and accumulated some 5000 years ago when sea levels were lower. In places, the original soil layer is preserved on top of the peat as a grey clay-rich layer (Figure 13.14a). The peat

(a)

(b)

Figure 13.14 Peat deposits, Wales. (a) Peat layer with rootlets and plant debris exposed on a beach. It is overlain by an *in situ* clay-rich, soil horizon, which also contains plant fragments, Borth. (b) Glacial diamict displaying soil development beneath an old planar land surface, which is overlain by a thin peat layer, Anglesey.

records a stage in climate change since the last glaciers retreated and tundra conditions were established. For example, the upper part of a glacial diamict exposed in a low sea cliff in Wales displays the formation of a soil profile beneath a planar erosion surface that is covered by a thin layer of peat (Figure 13.14b).

Such peat deposits provide evidence for global warming, and the buried forest in Wales provides proof of sea level rise over the last several thousands of years. Engineering projects to mitigate against sea level rise are being instigated, for example the construction of coastal defences along the coast of Wales (Figure 2.7b).

CASE STUDY 13.1   ALLUVIAL SEDIMENTS: BRIDGE FOUNDATIONS, PADMA RIVER, BANGLADESH

[23°28.6′N 90°16.0′E]

N.R. Wightman
Aquaterra Consultants

The proposed 6.15 km Padma Multipurpose Bridge will span 150 m between piers from the Mawa to the Janjira riverbanks, crossing one of the world's largest rivers, 35 km south-west of Dhaka City, Bangladesh (Figure CS13.1.1). The Mawa river bank is an old shoreline (Figure CS13.1.2) with transported soils having reclaimed ground 80 km to the south at the current coastline of the Bay of Bengal.

The annual sediment loading and deposition rate is over 2 billion tonnes of fine soils. The effects of river scour is significant, and at peak flow, the Padma River flow rate is over 125,000 m³/s with scour depths estimated to exceed 65 m with riverbanks prone to total erosion (Figure CS13.1.3). The bridge is to be supported on driven friction piles of 3 m diameter with toe levels of about −120 m depth. The location is in a seismically active area between two active east–west thrust faults and with earthquake tremors occurring frequently.

GEOLOGY

The geology comprises deep soil stratifications extending over 2 km in depth, which is underlain by limestone. The Padma Bridge, therefore, would require friction piles to support the bridge

Figure CS13.1.1 Graphic representation of the proposed bridge over the Padma River.

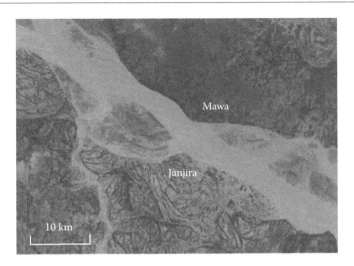

Figure CS13.1.2 Proposed location of the bridge across the Padma River, Google Earth, Landsat image. The bridge will be aligned north to south between Mawa and Janjira.

Figure CS13.1.3 Layered alluvium on the banks of the Padma River.

deck (Silva et al., 2010a,b), and the foundations would need to be located below the scour level of the river. The ground investigation provided original data and observations from geology at a deeper level than previously examined. The geological map of Bangladesh (Figure CS13.1.4) indicated estuarine sediments at the surface with no solid geology, which was confirmed by this investigation (Figure CS13.1.5). The borehole samples indicated loose- to medium-dense fine sands and silts to about −70 m (Figure CS13.1.6a). At this level, an unconformity was reached with an abrupt change to a layer, about 10 m thick, between −70 and −80 m below datum level

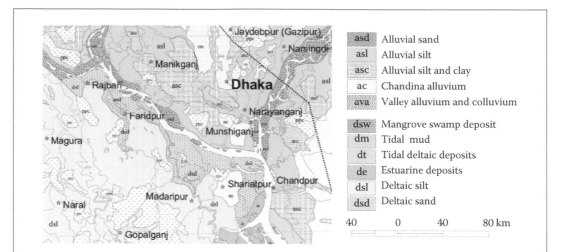

Figure CS13.1.4 Geological map of the Dhaka region. (From Bangladesh Geological Survey, 1:1,000,000 Geological Map of Bangladesh, 1990.)

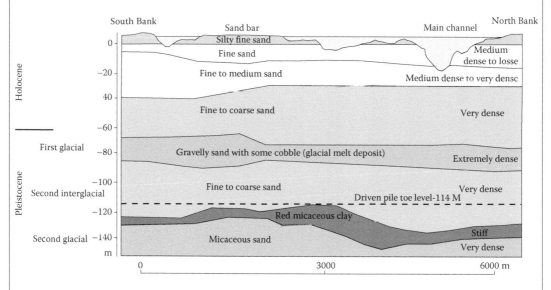

Figure CS13.1.5 Cross-section of the Quaternary deposits of the Padma River along the alignment of the proposed bridge.

of extremely dense, coarse sand with fine- to medium-sized, rounded gravel, which was deposited during rapid flow conditions from glacial meltwaters (Figure CS13.1.5). One medium-sized gravel showed signs of striations from glacial activity. The high density was the result of the thrust faulting causing vibratory compaction. The soils contained a range of abraded minerals, including mica, quartz, corundum and spinel. A very dense, grey and red clay layer with iron-stained mottling (Figure CS13.1.6b and c) was intersected at around −120 m below datum level, which formed during a previous interglacial period.

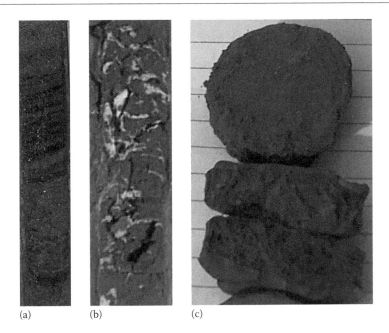

(a)   (b)   (c)

Figure CS13.1.6 Vibrocore samples of alluvium. Diameter of samples 6 cm. (a) Finely laminated, medium-dense, light-grey fine sand and dark-grey silt (−20 m). (b) Stiff, bioturbated, red clay with light-grey mottles (−125 m). (c) Very dense, grey clay with iron-stained mottling (−124 m).

ENGINEERING CONSIDERATIONS

- River sediments between coarse silt and fine sands with high mica content primarily from erosion of Himalayan micaceous schists.
- Extremely dense zone below −65 m depth with gravel and cobbles, compacted by recent thrust faulting.
- Highly micaceous red stiff clay and silt encountered below sand at a depth of around −120 m.
- The presence of hard, abrasive corundum (aluminium trioxide) within the alluvium resulted in rapid abrasion of the ground investigation cutting tools.

CASE STUDY 13.2 GLACIAL DEPOSITS: LANDSLIDE, ST DOGMAELS, WALES

[52°4.7′N 4°41.0′W]

Serious ground movements occurred across farmland above the village of St Dogmaels, West Wales (Figure CS13.2.1), in February 1994 after heavy rains. Scarps up to 3 m high developed across fields at an elevation of 120 mOD lower down the slope in which 25 houses sustained severe damage and some had to be abandoned. Consideration was given for the evacuation of the whole village (5000 people), but in the event, it was decided to immediately install a slope movement monitoring system across the landslide and instigate a forensic investigation, which included geological and geomorphological surveying, boreholes, trail pitting and installation of piezometres. A study of aerial photographs indicated that the landslide was the result of reactivation of much earlier failure, probably in postglacial times.

GEOLOGY

Distribution of bedrock (sandstone and shale), glacial deposits and landslide materials exposed in the St Dogmaels area is illustrated in Figure CS13.2.2. The glacial sequence (Figure CS13.2.3) overlies bedrock and, in places, a thin layer of preglacial silty sands and gravels (hill slope deposits); a 25 m thick unit of finely laminated silt and clay (glacio-lacustrine deposits), which includes a thin sand and gravel layer; and an uppermost glacial diamict complex composed of angular rock fragments set in a silty clay matrix. The landslide is largely composed of the glacio-lacustrine sediments, which are not exposed at the surface.

   The glacial deposits that cover large tracts of Britain were accumulated mainly during the last ice age (c. 20,000 years ago) when glaciers advanced to the south coast of Wales. At that time, an

Figure CS13.2.1 St Dogmaels landslide viewed from the north. The back scarp of the landslide is located close to farm at the right of the photograph and extends downslope as far as the houses in the higher part of the village.

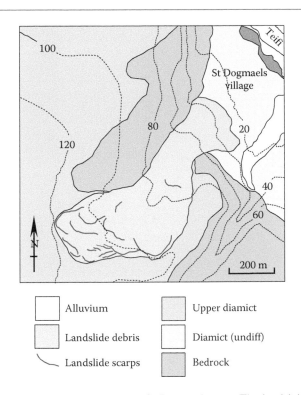

Figure CS13.2.2 Simplified geological map of the St Dogmaels area. The landslide material delineates the postglacial extent of the movement, whereas the landslide scarps are recent.

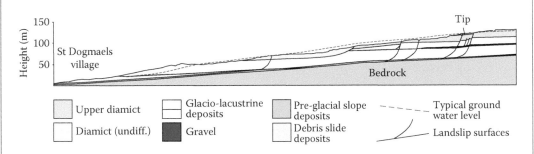

Figure CS13.2.3 Schematic cross-section of the landslide. The glacio-lacustrine deposits are divided into a lower and upper unit, separated by a thin gravel layer. No vertical exaggeration. (After Maddison, J.D. et al., 1999.)

ice sheet flowed southwards along the Irish Sea and dammed the Teifi Valley (Figure CS13.2.4a), thereby creating a deep glacial lake (Charlesworth, 1929; Fletcher and Siddle, 1998) that rose to over 200 m above present sea level and extended tens of kilometres inland (Hambrey et al., 2001). A wide variety of sediments were deposited in the lake in relation to the proximity of the glacier front (Figure CS13.2.4b). Mainly deltaic sands and gravels accumulated near the shoreline

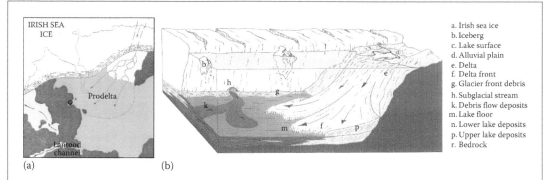

a. Irish sea ice
b. Iceberg
c. Lake surface
d. Alluvial plain
e. Delta
f. Delta front
g. Glacier front debris
h. Subglacial stream
k. Debris flow deposits
m. Lake floor
n. Lower lake deposits
p. Upper lake deposits
r. Bedrock

(a)   (b)

Figure CS13.2.4 Environments of deposition of the glacial deposits in the St Dogmaels area. (a) Location of St Dogmaels (red dot) relative to the shoreline of the former lake dammed by the Irish Sea ice sheet. (b) Vertically exaggerated schematic diagram of the glacial deposits within the ice-dammed lake.

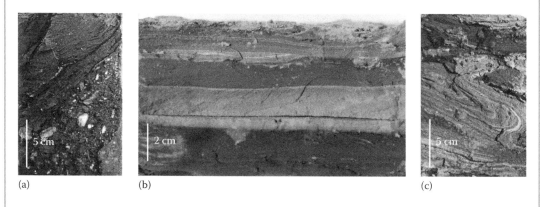

(a)   (b)   (c)

Figure CS13.2.5 Glacial deposits from the landslide site investigation. Mazier samples. (a) Graded, outwash, silty gravels overlying slightly laminated silty clay. (b) Close-up of laminated silts interbedded with faintly colour-banded clays. (c) Inclined, isoclinal slump fold in laminated silts.

(Figure CS13.2.5a), whereas the deeper parts of the lake were filled with finely laminated clays and silts (Figure CS13.2.5b). As the glaciers continued their advance, subglacial rivers discharged diamicts into the lake.

The original landslide developed in postglacial times when the retreat of the Irish Sea ice resulted in the rapid emptying of the ice-dammed lake and erosion of most of the lake sediments. However, in the steep tributary valleys, some of the soft lake sediments remained, and these experienced downslope slumping (Figure CS13.2.5c) and multiple regressive failures. Recent reactivation of the landslide due mainly to artesian water pressures and a waste dumping comprised non-circular failures related to the main aquifers within the glacial sequence (Figure CS13.2.3).

ENGINEERING CONSIDERATIONS

- A comprehensive model of the glacial environment was required to fully understand the distribution of the sediments and the formation of the landslide.
- Photogeological surveys of the area would have identified the glacial landslip.
- Artesian groundwater pressures are present across the site, and water flow is concentrated in a basal preglacial hillslope deposit and a thin gravel layer within the lacustrine sediments.
- Installation of well drainage has arrested landslide movements.

CASE STUDY 13.3   LOESS: ENERGY INFRASTRUCTURE DEVELOPMENT, SOUTH OF THE RIVER URAL, NORTH-WESTERN KAZAKHSTAN

[51°21.2′N 53°12.5′E]

C. Roohnavaz, E.J.F. Russell and H.F. Taylor
Mott MacDonald Ltd., Croydon, U.K.

A large expansion to one of the world's largest processing facilities (Figure CS13.3.1) is located on unsaturated metastable loess soils that occur within the Pliocene to recent alluvial deposits of the River Ural (Figure CS13.3.2). The construction of the facility required extensive grading and excavations producing hundreds of thousands of cubic metres of loess considered unsuitable for reuse as earthwork. An understanding of loess structure, mineralogy and collapse mechanism was necessary to engineer a stable loess for reuse, thereby significantly increasing the sustainable aspects of the development (Roohnavaz et al., 2010). Loess microstructure and its mineralogy were identified through thin sections and petrographic analysis.

Figure CS13.3.1 View of the subdued loess topography and processing facility.

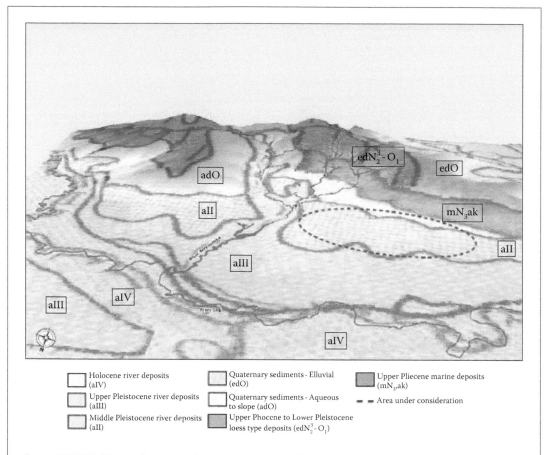

Figure CS13.3.2 Three-dimensional topography and drift geology.

GEOLOGY

Loess accumulates by aeolian processes in a continental climate characterized by cold winters and hot dry summers. Under such a climate, net upward flux of water results in gradual drying, cracking and desiccation of the soil mass. The loess at the site is predominantly stiff to hard silt and silty clay with fine, horizontal and inclined laminations (Figure CS13.3.3). Exposure of soils above river level, during formation, is evidenced by numerous rootlets and rootlet tracks at some horizons and occasional buried (fossil) topsoil layers. During periods of exposure, including the present day, it is considered that aeolian processes have dominated, increasing evaporation and rapidly depositing layers of windblown material. Increased evaporation has resulted in the formation of a desiccated surface layer that is drier and stiffer than the underlying soil and contains subvertical desiccation cracks. Due to the loess deposition mode, it is considered that clay 'bridges' connecting silt particles have maintained the open structure (Rogers et al., 1994). Mobilization and redeposition of carbonate by percolating waters is evidenced by nodules within the strata and is particularly evident near the surface (Figure CS13.3.4). The presence of these nodules may suggest that carbonate bonding has formed early during deposition, helping to maintain the open structure. It is likely that these characteristics combined with the partially saturated condition of the soils have resulted in the soils being susceptible to collapse when loaded and wetted.

Figure CS13.3.3 Laminated loess from borehole (depth 12 m, diameter of sample 12 cm).

Figure CS13.3.4 Trial pit in loess mass showing carbonate nodules and desiccation cracks.

Loess thin sections identified quartz, feldspar and calcite and small amounts of iron, organic matter and glauconite in a matrix of clay compounds. The petrographic analyzes indicated a silt-rich matrix with large voids (Figure CS13.3.5). The clay fraction in all samples contained the sheet silicate mineral smectite, which would be expected to be moisture sensitive and capable of undergoing volume change when wetted or dried.

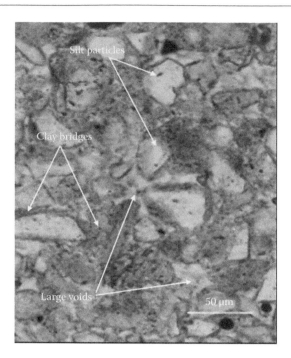

Figure CS13.3.5 Thin section showing loess silt-rich matrix, clay bridges and voids.

In these soils, particles of silt are cemented together by aggregations of clay compounds and carbonate 'bridges' forming a porous structure which is hard when dry but liable to rapidly change state and collapse under wetting and pressure. Testing of the loess under natural moisture content and submerged conditions and with measurements of suction indicated a dual 'hydrodynamic' process leading to its collapse. The tests indicated that the strong loess structure when dry had two components: bonding between particles due to salts and carbonates and suction due to its porous particle arrangements. Loess collapse under loading and wetting occurs when bonding between the particles is weakened and suction removed causing the particles to move into a closer pack reducing the void spaces. A series of compaction tests on loess at varying moisture content and compactive effort aided with suction measurements were used to engineer a stable loess mass which showed significant reduction in suction.

ENGINEERING CONSIDERATIONS

- The distribution of collapsible soils is underrepresented by the regional geological maps.
- The loess in its natural state when excavated is unsuitable for reuse in earthworks.
- Aggregations of clay compounds and carbonate bridges connecting the silt particles form an open and porous structure, which is hard when dry but liable to change state rapidly and collapse under wetting.
- Reworking and compaction of the loess reduces suction thereby allowing the material to be used as a sustainable resource.

References

Abweny, M.S. 2009. *The Geology of Tallat Al Bustana and Wadi Al Mirba' Areas, Map Sheets No. 375-V and 385-V.* The Hashemite Kingdom of Jordan Natural Resources Authority, 70, Amman, Jordan.

Anderson, D.L. 1989. *Theory of the Earth.* Blackwell Scientific, Oxford, U.K.

Arup. 1998. West Rail Yuen Long Section, geotechnical data report. West Rail Detailed Design Contract DD-20. Ove Arup & Partners Ltd., for Kowloon-Canton Railway Corporation (KCRC), Hong Kong, (unpublished).

Arup. 2001. Stonecutters bridge site investigation and desk study report, Vol. 2. Agreement No. CE61/2000. Ove Arup & Partners Ltd., for Highways Department, Hong Kong SAR Government, Hong Kong, (unpublished).

Bangladesh Geological Survey. 1990. 1:1,000,000 Geological Map of Bangladesh, Government of Bangladesh.

Barton, N. 2002. Some new Q-value correlations to assist in site characterization and tunnel design. *International Journal of Rock Mechanics and Mining Sciences*, 39, 185–216.

Beard, L.S., Anderson, R.E., Block, D.L., Bohannon, R.G., Brady, R.J., Castor, S.B. and Duebendorfer, E.M. et al. 2007. Preliminary geologic map of the Lake Mead 30′ × 60′ quadrangle, Clark County, Nevada, and Mohave County, Arizona: U.S. Geological Survey Open-File Report 2007–1010, 109pp, 3 plates, scale 1:100,000.

Bennison, G.M. 1985. *An Introduction to Geological Structures and Map.* Edward Arnold, London, U.K.

Bews, I. 2012. Stabilising Glandyfi. *Geoscientist*, 22(1), 19–21.

Bhattacharya, J.P. and Walker, R.G. 1992. Deltas. In: Walker, R.G. and James, N.P. (eds.), *Facies Models – Response to Sea Level Change.* Geological Association of Canada, St. John's, Newfoundland, Canada, pp. 157–175.

Bieniawski, Z.T. 1989. *Engineering Rock Mass Classifications: A Complete Manual for Engineers and Geologists in Mining, Civil, and Petroleum Engineering.* Wiley-Interscience, New York, pp. 40–47.

Blythe, F.G.H. and de Freitas, M.H. 1984. *A Geology for Engineers*, 7th edn. Edward Arnold, London, U.K., 325pp.

Bock, H. 2006. Common ground in engineering geology, soil mechanics and rock mechanics: Past, present and future. *Bulletin of Engineering Geology and the Environment* (IAEG), 65, 209–216.

Boggs, S. 2009. *Petrology of Sedimentary Rocks*, 2nd edn. Cambridge University Press, U.K.

Boulter, C.A. 1989. *Four Dimensional Analysis of Geological Maps.* John Wiley & Sons, Chichester, U.K.

Brathwaite, C.J.R. 2005. *Carbonate Sediments and Rocks: A Manual for Earth Scientists and Engineers.* Whittles, University of Michigan, MI.

British Geological Survey. 1962. Bristol District, One-inch Geological Sheets 264 and 280, Natural Environment Research Council, U.K.

British Geological Survey. 1989. Aberystwyth 1:50,000 Sheet 163, Solid and Drift Edition, Natural Environment Research Council, U.K.

British Standards Institution. 1999. BS 5930:1999. *Code of Practice for Site Investigation*, 206pp.

British Standards Institution. 2002. EN ISO 14688-1. *Geotechnical Investigation and Testing – Identification and Classification of Soil – Part 1: Identification and Description.*

British Standards Institution. 2003. EN ISO 14689-1. *Geotechnical Investigation and Testing – Identification and Classification of Rock – Part 1: Identification and Description.*

Brookfield, M.E. 1992. Eolian deposits. In: Walker, R.G. and James, N.P. (eds.), *Facies Models – Response to Sea Level Change.* Geological Association of Canada, St. John's, Newfoundland, Canada, pp. 143–156.

Brunsden, D., Doornkamp, J.C., Fookes, P.G., Jones, D.K.C. and Kelly, J.M.H. 1975. Large scale geomorphological mapping and highway engineering design. *Quarterly Journal of Engineering Geology*, 8, 227–253.

Brunsden, D., Jones, D.K.C., Martin, R.P. and Doornkamp, J.C. 1981. The geomorphological character of part of the Low Himalaya of Eastern Nepal. *Zeitschrift fur Geomorphologie*, 27, 25–72.

Bureau of Geology and Mineral Resources of Guangdong Province. 1988. *Regional Geology of Guangdong Province.* Geological Memoirs Series 1, 941pp, Government of Guangdong Province, P.R. China.

Chan, Y.C. and Pun, W.K. 1994. Karst morphology of foundation design. GEO report 32, Geotechnical Engineering Office, Civil Engineering Department, Hong Kong, People's Republic of China, 90pp.

Charlesworth, J.K. 1929. The south Wales end-moraine. *Quaternary Journal of the Geological Society of London*, 85, 335–358.

Collinson, J.D., Mountney, N. and Thompson, D.B. 2006. *Sedimentary Structures*, 3rd edn. Terra Publishing, Dunedin Press, U.K.

Dugar, S. 2013. *Effect of Seismic Hazards on Low Cost Mountain Roads.* MSc thesis, Institute of Hazard, Risk and Resilience, Durham University, Durham, U.K., and Poster Presentation at the Himalayan Karakorum Tibet Workshop and International Symposium on Tibetan Plateau, Tubingen, Germany.

Eckardt, F.D. and Spiro, B. 1999. The origin of sulphur in gypsum and dissolved sulphate in the Central Namib Desert, Namibia. *Sedimentary Geology*, 123, 255–273.

Eyles, N. and Eyles, C.H. 1992. Glacial depositional systems. In: Walker, R.G. and James, N.P. (eds.), *Facies Models – Response to Sea Level Change.* Geological Association of Canada, St. John's, Newfoundland, Canada, pp. 73–100.

Faure, G. 1986. *Principles of Isotope Geology*, 2nd edn. John Wiley & Sons, New York.

Fettes, D.J. and Desmons, J. (eds.). 2011. *Metamorphic Rocks – A Classification and Glossary of Terms: Recommendations of the International Union of Geological Sciences Subcommission on the Systematics of Metamorphic Rocks.* Cambridge University Press, Cambridge, U.K.

Fisher, R.V. and Schminke, H.-U. 1984. *Pyroclastic Rocks.* Springer-Verlag, Berlin, Germany.

Fletcher, C.J.N. 1978. *Structures in Folded Rocks – An Illustrated Guidebook to Structural Geology* (English, Spanish and Korean versions). Ministry of Overseas Development, United Kingdom Government, London, U.K.

Fletcher, C.J.N. 2004. *Geology of Site Investigation Boreholes from Hong Kong.* Applied Geoscience Centre, Hong Kong Construction Association and Association of Geotechnical Specialists, Hong Kong, 132pp.

Fletcher, C.J.N. 2006. Importance of comprehensive rock and soil description in ground modelling. *Proceedings of the Hong Kong Soils and Rocks Symposium*, Hong Kong, People's Republic of China, pp. 17–28.

Fletcher, C.J.N., Collar, F.A. and Lai, M.C.W. 2000b. Marine magnetic survey of Hong Kong: Results, interpretation and significance. In: Page, A. and Reels, S.J. (eds.), *The Urban Geology of Hong Kong*, Vol. 6. Geological Society of Hong Kong, Hong Kong, pp. 179–188.

Fletcher, C.J.N., Massey, C.I., Williamson, S.J. and Parry, S. 2002. Importance of bedrock and regolith mapping for natural terrain hazard studies: An example from the Tsing Shan area, Hong Kong. *Natural Terrain – A Constraint to Development*, Institution of Mining and Metallurgy, Hong Kong.

Fletcher, C.J.N., Pang, V.P.Y. and Yim, K.P. 2003. Preliminary geological modelling and foundation appraisal for project planning of development sites in Tin Shiu Wai, Northwestern New Territories. *Proceedings of Foundation Practice in Hong Kong Symposium*, Hong Kong, P.R. China.

Fletcher, C.J.N. and Siddle, H.J. 1998. The development of glacial Lake Teifi – Evidence for lake-level fluctuations at the margin of the Irish Sea ice sheet. *Journal of the Geological Society, London*, 155, 389–400.

Fletcher, C.J.N., Wightman, N.R. and Goodwin, C.R. 2000a. Karst related deposits beneath Tung Chung New Town: Implications for deep foundations. *Engineering Geology HK 2000*. Institution of Mining and Metallurgy, Hong Kong, pp. 139–150.

Fookes, P.G. 1997. Geology for engineers: The geological model, prediction and performance (First Glossop Lecture). *Quarterly Journal of Engineering Geology*, 30(4), 293–424.

Fookes, P.G., Baynes, F.J. and Hutchinson, J.N. 2000. Total geological history: A model approach to anticipation, observation and understanding of site conditions. In *Proceedings of the International Conference on Geotechnical and Geological Engineering. GeoEng 2000*, Melbourne Technomic Publishing, Basel, Switzerland.

Fookes, P.G., French, W.J. and Rice, S.M.M. 1985. The influence of ground and groundwater geochemistry on construction in the Middle East. *Quarterly Journal of Engineering Geology*, 18, 101–128.

Fookes, P.G. and Horswill, L.P. 1970. Discussion on engineering grade zones. In *Proceedings of the In Situ Investigations in Soils and Rocks*, London, U.K., pp. 53–57.

Fookes, P.G. and Marsh, A.H. 1981. Some characteristics of construction materials in the low to moderate metamorphic grade rocks of the Lower Himalayas of East Nepal: 1. Occurrence and geological features. 2. Engineering characteristics. *Proceedings of the Institution of Civil Engineers, Part 1*, 70, 123–162.

Fookes, P.G., Sweeney, M., Manby, C.N.D. and Martin, R.P. 1985. Geological and geotechnical engineering aspects of low-cost roads in mountainous terrain. *Engineering Geology*, 21, 1–152.

Frost, D.V. 1992. *Geology of Yuen Long*. Hong Kong Geological Survey Sheet Report No. 1. Geotechnical Engineering Office, Civil Engineering Department, Hong Kong, People's Republic of China.

Fry, N. 1984. *The Field Description of Metamorphic Rocks*. Open University Press, U.K.

Fyfe, J.A., Shaw, R., Campbell, S.D.G., Lai, K.W. and Kirk, P.A. 2000. *The Quaternary Geology of Hong Kong*. Hong Kong Geological Survey, Geotechnical Engineering Office, Civil Engineering Department, The Government of the Hong Kong SAR, Hong Kong, People's Republic of China, 209pp.

Geological Society. 1991. *Field Description of Metamorphic Rocks*. In *Geological Society Professional Handbook*. Geological Society of London, London, U.K.

Geological Society. 1997. Tropical residual soils. In: Fookes, P.G. (ed.), *Geological Society Professional Handbook*. Geological Society of London, London, U.K.

Geological Society of Canada. 1992. In: Walker, R.G. and James, N.P. (eds.), *Facies Models – Response to Sea Level Change*. Geological Association of Canada, St. John's, Newfoundland, Canada.

Geological Survey of Bangladesh. 1990. Geological map of Bangladesh 1,000,000 scale. Digitally compiled by U.S. Geological Survey, 2001.

Geotechnical Control Office. 1987. *Guide to Site Investigation (Geoguide 2)*. Hong Kong, 359pp.

Geotechnical Control Office. 1988a. *Guide to Rock and Soil Descriptions (Geoguide 3)*. Geotechnical Control Office, Hong Kong, 189pp.

Geotechnical Control Office. 1988b. *Yuen Long*. Hong Kong Geological Survey Sheet 6, Solid and Superficial Geology, 1:20,000 Series HGM20, Geotechnical Control Office, Hong Kong.

Geotechnical Control Office. 1989. Sheet 6-NW-B Yuen Long, 1:5,000 scale. Superficial Geology, Hong Kong.

Geotechnical Engineering Office. 1987. *Hong Kong South & Lamma Island*. Hong Kong Geological Survey Sheet 15. Solid and Superficial Geology. 1:20,000 Series HGM20. Geotechnical Engineering Office, Hong Kong.

Geotechnical Engineering Office. 1996. Report on the Shum Wan Road landslide of 13 August 1995 – Volume 2. Findings of the landslide investigation. Geotechnical Engineering Office, Hong Kong.

Geotechnical Engineering Office. 2000. *Geological Map of Hong Kong*, Millennium Edition. 1:200,000 scale. Civil Engineering and Development Department, Geotechnical Engineering Office, Hong Kong.

Geotechnical Engineering Office. 2004. *Guidelines on Geomorphological Mapping for Natural Terrain Hazard Studies. Technical Guidance Note 22.* Civil Engineering and Development Department, Geotechnical Engineering Office, Hong Kong.

Geotechnical Engineering Office. 2006. *Foundation Design and Construction.* Civil Engineering and Development Department, Geotechnical Engineering Office, Hong Kong, 348pp.

Geotechnical Engineering Office. 2007. *Engineering Practice in Hong Kong*, GEO Publication No. 1/2007. Civil Engineering and Development Department, Geotechnical Engineering Office, Hong Kong, 278pp.

Hall, A. 1996. *Igneous Petrology.* Longman Group Ltd., Harlow, U.K.

Hambrey, M.J., Davies, J.R., Glasser, N.F., Waters, R.A., Dowdeswell, J.A., Wilby, P.R., Wilson, D. and Etienne, J.L. 2001. Devensian glacigenic sedimentation and landscape evolution in the Cardigan area of southwest Wales. *Journal of Quaternary Science*, 16, 455–482.

Harland, W.B., Cox, A.V., Llewellyn, P.G., Picton, C.A.G., Smith, A.G. and Walters, R. 1982. *A Geological Time Scale.* Cambridge University Press, Cambridge, U.K.

Hearn, G.J. 2002. Engineering geomorphology for road design in unstable mountainous areas: Lessons learned after 25 years in Nepal. *Quarterly Journal of Engineering Geology and Hydrogeology*, 35, 143–154.

Hencher, S. 1987. Implications of joints and structures for slope stability. In: Anderson, M.G. and Richards, K.S. (eds.), *Slope Stability – Geotechnical Engineering and Geomorphology.* John Wiley & Sons, U.K., pp. 145–186.

Hencher, S. 2012. *Practical Engineering Geology.* Spon Press, London, U.K., 450pp.

Hencher, S. and Daughton, G. 2000. Anticipating geological problems. In: Page, A. and Reels, S.L. (eds.), *The Urban Geology of Hong Kong*, Vol. 6. Geological Society of Hong Kong, Hong Kong, pp. 89–98.

Henson, R. 2006. *The Rough Guide to Climate Change.* Rough Guides, London, U.K., 352pp.

Howell, J. 1999. *Roadside Bio-Engineering – Reference Manual and Site Handbook.* Department of Roads, His Majesty's Government of Nepal, Nepal.

Hutchinson, C.S. 1989. *Geological Evolution of South-East Asia.* Oxford University Press, Oxford, U.K.

International Society for Rock Mechanics. 1978. Suggested methods for quantitative description of discontinuities in rock masses. *International Journal of Rock Mechanics and Mining Sciences & Geomechanics Abstracts*, 15, 319–368.

Jennings, J.N. 1985. *Karst Geomorphology.* Blackwell, Oxford, U.K.

Kendall, A.C. 1992. Evaporites. In: Walker, R.G. and James, N.P. (eds.), *Facies Models – Response to Sea Level Change.* Geological Association of Canada, St. John's, Newfoundland, Canada, pp. 375–409.

Kirk, P.A. 2000. Adverse ground conditions at Tung Chung New Town. In: Page, A. and Reels, S.L. (eds.), *The Urban Geology of Hong Kong*, Vol. 6. Geological Society of Hong Kong, pp. 89–98.

Kirk, P.A., Campbell, S.D.G., Fletcher, C.J.N. and Merriman, R.J. 1997. The significance of primary volcanic fabrics and clay distribution in landslides in Hong Kong. *Journal of the Geological Society of London*, U.K. 154, 1009–1019.

Knill, J.I. 2002. Core values: The first Hans Cloos Lecture. In *Proceedings of the Ninth AEG Congress*, Durban, South Africa, pp. 1–45.

Le Maitre, R.W. (ed.). 1989. *A Classification of Igneous Rocks and Glossary of Terms. Recommendations of the International Union of Geological Sciences on the Systematics of Igneous Rocks.* Blackwell Scientific Publications, Oxford, U.K., 193pp.

Litherland, M., Annells, R.N., Appleton, J.D., Berrange, J.P., Bloomfield, K., Fletcher, C.J.N. and Hawkins, M.P. et al. 1986. *The Geology and Mineral Resources of the Bolivian Precambrian Shield.* Overseas Memoir of the British Geological Survey, No. 9, 152pp.

Mackay, A.D., Wightman, N.R. and Houghton, A. 2013. Site investigation approaches for the proposed Ukhuu Khudag to Gashuun Sukait Railway, South Gobi Desert, Mongolia. *Proceedings of ISGSR 2013*, Hong Kong, People's Republic of China.

Maddison, J.D., Siddle, H.J. and Fletcher, C.J.N. 1999. Investigation and remediation of a major landslide in glacial lake deposits at St Dogmaels, Pembrokeshire. In *Proceedings of the Eighth International Symposium on Landslides*, Cardiff, U.K.

Malone, A.W., Hansen, A., Hencher, S.R. and Fletcher, C.J.N. 2008. Post-failure movements of a large slow rock slide in schist near Po Selim, Malaysia. In *Proceedings of the 10th International Symposium on Landslides and Engineered Slopes*, Xi'an, China, pp. 457–461.

Maltman, A. 1998. *Geological Maps: An Introduction*, 2nd edn. John Wiley & Sons, Chichester, U.K.

Marinos, P. and Hoek, E. 2000. GSI – A geologically friendly tool for rock mass strength estimation. In *Proceedings GeoEng2000 Conference*, Melbourne, Victoria, Australia, pp. 1422–1442.

Martin, R.P. 2001. The design of remedial works to the Dharan-Dhankuta Road, East Nepal. In: Griffiths, J.S. (ed.), *Land Surface Evaluation for Engineering Practice*. Engineering Geology Special Publication, 18. Geological Society of London, London, U.K, pp. 197–204.

Martin, R.P. and Hencher, S.R. 1986. Principles for description and classification of weathered rocks for engineering purposes. In: Hawkins, A.B. (ed.), *Site Investigation Practice: Assessing BS 5930*. Engineering Geology Special Publication, 2. Geological Society of London, London, U.K, pp. 299–308.

Martinson, D.G., Pisias, N.G., Hays, J.D., Imbrie, J., Moore, T.C. and Shackleton, N.J. 1987. Age dating and the orbital theory of the Ice Ages: Development of a high resolution 0 to 300,000 year chronostratigraphy. *Quaternary Research*, 27, 1–29.

McClay, K. 1991. *The Mapping of Geological Structures*. Butler and Tanner Ltd., London, U.K.

McPhie, J., Doyle, M. and Allen, R. 1993. *Volcanic Textures: A Guide to the Interpretation of Textures in Volcanic Rocks*. Centre for Ore Deposits and Exploration Studies, University of Tasmania, New Zealand, 198pp.

Miall, A.D. 1992. Alluvial deposits. In: Walker, R.G. and James, N.P. (eds.), *Facies Models – Response to Sea Level Change*. Geological Association of Canada, St. John's, Newfoundland, Canada, pp. 119–142.

Mitchell, J.M. 1986. Foundations for the Pan Pacific Hotel on pinnacled and cavernous limestone. In: Chan, S.F. (ed.), *Foundation Problems in Limestone Areas of Peninsular Malaysia*. The Geotechnical Engineering Division of the Institution of Engineers Malaysia, Malaysia, pp. 37–55.

Mongolian Geological Survey. 2010. 1:5,000,000-scale Geological Map of Mongolia, Government of Mongolia.

Nichols, G. 2009. *Sedimentology and Stratigraphy*, 2nd edn. Wiley-Blackwell, Oxford, U.K.

Norbury, D. 2010. *Soil and Rock Description in Engineering Practice*. CRC Press, Florida, U.S.A., 288pp.

North American Stratigraphic Code. 2005. AAPG Bulletin 89/11, U.S.A, pp. 1547–1591.

Nowell, D. 2014. Italian geological maps. *Geoscientist*, 24(4), 10–15.

Parry, S. and Ruse, M.E. 2002. The importance of geomorphology for natural terrain hazard studies. In *Proceedings of the Conference Natural Terrain – A Constraint to Development*. Institution of Mining and Metallurgy, Hong Kong Branch, Hong Kong, pp. 89–100.

Peck, D.L. 2002. *Geological Map of the Yosemite Quadrangle, Central Sierra Nevada, California*. Geologic Investigation Series Map I-2751, United States Geological Survey, U.S.A.

Pinches, G., Tosen, R. and Thompson, T. 2000. The contribution of geology to the engineering of Hong Kong International Airport. In: Page, A. and Reels, S.I. (eds.), *Urban Geology of Hong Kong*, Vol. 6. Geological Society of Hong Kong, Hong Kong, pp. 21–42.

Plant, G.W., Covil, C.S. and Hughes, R.A. (eds.). 1998. *Site Preparation for the New Hong Kong International Airport*. Thomas Telford, U.K., 576pp.

Ramsey, J.G. and Huber, M.I. 1987. *The Techniques of Modern Structural Geology. Folds and Fractures*, Vol. 2. Academic Press, London, U.K., 462pp.

Rogers, C.D.F., Dijkstra, T.A. and Smalley, I.J. 1994. Hydroconsolidation and subsidence of loess: Studies from China, Russia, North America and Europe. *Engineering Geology*, 37(2), 83–113.

Roohnavaz, C., Russell, E.J.F. and Taylor, H.F. 2010. Unsaturated loessial soils: A sustainable solution for earthwork. In *Geotechnical Engineering, Proceedings of the Institution of Civil Engineers*.

Ruxton, B.P. and Berry, L. 1957. Weathering of granite and associated erosional features in Hong Kong. *Bulletin of Geological Society of America*, 68, 1263–1292.

Saunders, M.K. and Fookes, P.G. 1970. A review of the relationship of rock weathering and climate and its significance to foundation engineering. *Engineering Geology*, 4, 289–325.

Sewell, R.J., Campbell, S.D.G., Fletcher, C.J.N., Lai, K.W. and Kirk, P.A. 2000. *The Pre-Quaternary Geology of Hong Kong*. Geotechnical Engineering Office, Civil Engineering and Development Department, Hong Kong, 180p.

Sewell, R.J. and Fletcher, C.J.N. 2002. Regolith mapping in Hong Kong and its application to studies of landslide susceptibility. *Natural Terrain – A Constraint to Development?* Institution of Mining and Metallurgy, Hong Kong, pp. 183–196.

Shackelton, N.J. 1987. Oxygen isotopes, ice volume and sea level. *Quaternary Science Reviews*, 6, 183–190.

Silva, S.D., Wightman, N.R. and Kamruzzaman, M. 2010a. Geotechnical ground investigation of the Padma Main Bridge. In *Proceedings of the IABSE-JSCE Joint Conference on Advances in Bridge Engineering II*, Dhaka, Bangladesh, pp. 427–436.

Silva, S.D., Wightman, N. and Kamruzzaman, M. 2010b. Geotechnical ground investigation of the Padma Main Bridge. *Proceedings of the Bangladesh Geotechnical Conference 2010: Natural Hazards and Countermeasures in Geotechnical Engineering, ISSMGE, 2010*, Dhaka, Bangladesh, pp. 181–188.

Stephenson, S. and Merritt, J. 2006. *Skye, A Landscape Fashioned by Geology*. Scottish Natural Heritage, U.K., 22pp.

Stow, D.V.A. 2005. *Sedimentary Rocks in the Field: A Colour Guide*. Manson Publishing, CRC Press, U.K.

Summerfield, M.A. 1991. *Global Geomorphology*. Pearson, Prentice Hall, UK, 537pp.

Sweeting, M.M. 1972. *Karst Landforms*. MacMillan, London, U.K.

Thomas, L. 2012. *Coal Geology*, 2nd edn. Wiley-Blackwell, Oxford, U.K.

Tucker, M.E. 2001. *Sedimentary Petrology*. Blackwell Publishing, Oxford, U.K.

van der Pluijm, B.A. and Marshak, S. 2004. *Earth Structure, An Introduction to Structural Geology and Tectonics*, 2nd edn. Norton & Company, New York.

Vine, F.J. and Matthews, D.H. 1963. Magnetic anomalies over oceanic ridges. *Nature*, 199, 947–949.

Walker, R.G. 1992. Turbidites and submarine fans. In: Walker, R.G. and James, N. P (eds.), *Facies Models – Response to Sea Level Change*. Geological Association of Canada, Canada, pp. 239–263.

Waltham, A.C. 2009. *Foundations of Engineering Geology*, 3rd edn. Spon Press, London, U.K., 92pp.

Wightman, N.R., Hitchcock, B.K., Burdidge, H.T., Frappin, P. and Goodwin, C.R. 2001. Advances in site investigation practice for deep foundations, Tung Chung New Town, Lantau Island, Hong Kong. In *Proceedings of the International Conference on In-Situ Measurement of Soil Properties and Case Histories*, Bali, Indonesia, May 2002.

Wilson, D., Davies, J.R., Fletcher, C.J.N. and Smith, M. 1990. *Geology of the South Wales Coalfield, Part VI, the Country around Bridgend*. Memoir of the British Geological Survey, Sheet 261 and 262 (England and Wales), Natural Environment Research Council. U.K., 62pp.

Winchester, S. 2001. *The Map that Changed the World*. Penguin Books, London, U.K.

Yeap, E.B. 1986. Irregular topography of the subsurface carbonate bedrock in Kuala Lumpa, area. In: Chan, S.F. (ed.), *Foundation Problems in Limestone Areas of Peninsular Malaysia*. The Geotechnical Engineering Division of the Institution of Engineers Malaysia, pp. 1–12.

Yuen, K.M. 1990. Characteristics and petrography of the Carboniferous Yuen Long arble. Geological Society of Hong Kong, Hong Kong, 4, pp. 73–87.

Index